E. Deeson

M.Sc., F. Inst. P., F.C.P., F.R.A.S.

Technician physics

Level 2

Longman London and New York

Longman Group Limited
Longman House Burnt Mill, Harlow, Essex, England
CM20 2JE Associated companies throughout the world.

Published in the United States of America
by Longman Inc., New York

First published 1984

British Library Cataloguing in Publication Data

Deeson, Eric
  Technician physics, level 2. – (Longman
  technician series: mathematics and sciences)
  1. Physics
  I. Title
  530′.0246   QC21.2   80-41946

  ISBN 0-582-41585-3

Set in 10/11 pt Linotron 202 Times
Printed in Singapore by Selector Printing Co (Pte) Ltd

Longman Technician Series

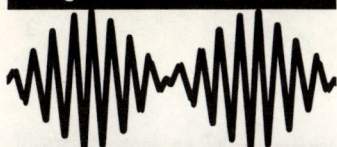

Mathematics and Sciences

Sector Editor:

D R Browning, B.Sc., F.R.S.C., C.Chem., A.R.T.C.S.
Principal Lecturer and Head of Chemistry, Bristol Polytechnic

Books already published in this sector of the series:

Technician mathematics Level 1   *J O Bird and A J C May*
Technician mathematics Level 2   *J O Bird and A J C May*
Technician mathematics Level 3   *J O Bird and A J C May*
Technician mathematics Levels 4 and 5   *J O Bird and A J C May*
Mathematics for science technicians Level 2   *J O Bird and A J C May*
Mathematics for electrical and telecommunications technicians
      Level 2   *J O Bird and A J C May*
Mathematics for electrical technicians Level 3   *J O Bird and A J C May*
Mathematics for electrical technicians Levels 4 and 5   *J O Bird and A J C May*
Calculus for technicians   *J O Bird and A J C May*
Statistics for technicians   *J O Bird and A J C May*
Physical sciences Level 1   *D R Browning and I McKenzie Smith*
Engineering science for technicians Level 1   *D R Browning and I McKenzie Smith*
Safety science for technicians   *W J Hackett and G P Robbins*
Fundamentals of chemistry   *J H J Peet*
Further studies in chemistry   *J H J Peet*
Technician chemistry Level 1   *J Brockington and P J Stamper*
Mathematics for scientific and technical students   *H G Davies and G A Hicks*
Mathematical formulae for TEC courses   *J O Bird and A J C May*
Science formulae for TEC courses   *D R Browning*
Organic chemistry for higher education   *J Brockington and P J Stamper*
Microprocessors and control   *J F A Thompson*
Cell biology for technicians level 2   *N A Thorpe*

Series Editor

D R Browning, B Sc, F.R.S.C., C.Chem, A.R.I.C.S.
Principal Lecturer... Head of Chemistry, University of ...

Books already published in this series

Contents

viii

Introduction

Physics for Technicians, Level 2, provides a clear and very comprehensive development of the material set out in the Technician Education Council Level 2 Physics Course.

Where I have felt it necessary, I have extended the objectives to include more background facts. The aim has been to develop a text which is coherent in itself, a more than adequate base for Level 3 work, and a good account of an important living field of knowledge.

In addition, the Level 1 objectives on which this work is based are revised briefly but with care. This should help those who have problems with that work, and those who, for any reason, have not studied TEC Level 1 physics before starting Level 2.

It also seemed important to write a book which can be followed by the student in isolation. This is part of the reason for the extension of the coverage below and beyond the TEC objectives list. It has also led to the unusually high content of illustrations, and the large number (over 450) of worked and unworked questions. A major problem faced by the student working in isolation is the lack of laboratory facilities. This book is not a practical text but physics is a practical subject. Indeed the 1981 version of the TEC Level 2 Physics specification explicitly lists a number of practical objectives. Details on these appear in the 'appropriate' sections in the book.

A big problem with questions in a textbook is that one can be tempted to work back from the answers. To make this less easy nearly all the unworked questions in this book have been given a random code number. The answers, comments and hints to unworked questions appear at the back of the book in numerical order. Thus at the start of Chapter 1, the first question is coded [100] and the second is [250]. When you read the answer to the first you will not see the answer to the second.

Throughout the preparation of the text, I have tried to make it as easy to read as I can. Science concepts can be fairly hard. While we must be precise when we discuss these, we should not hide them in unnecessarily difficult language. In recent years, there has been much research into the 'readability' of texts; I have tried to take account of the findings.

This book is one of a series published under the guidance of David Browning. I must record my great debt to David's work on the project. I have not found it common to have such a careful and knowledgeable editor. I must also thank my typist, Ann Adamson, for a good job done on a badly written and fairly tough manuscript.

Lastly I thank my family – Cynthia, Nicola, Rebecca and Richard – for their patience and understanding throughout the long pregnancy of this volume. I gladly dedicate the product to them.

Eric Deeson
Harborne, Birmingham 1983

Part I

Dynamics: forces and their effects on objects

Chapter 1

Motion

1.1 Speed

In the first part of this book we learn about **forces**. What are they? What do they do? How do they relate to other things?

We know two effects of forces. A force can make something move; a force can stop something moving.

So we first discuss motion. This is often described by '**speed**'.

The speed of an object is the distance moved in unit time.

Speed = distance moved/time taken
Unit: metre per second, m s^{-1}

People often use other units of speed in daily life; an instance is the kilometre per hour, km h^{-1}. In this book we work with SI units. The SI unit of speed is the metre per second (m s^{-1}).

In 1979 Sebastian Coe ran 1500 m in 3 minutes 32 seconds. What was his speed?   [100]

I walked 9 km in 2.5 h. What was my speed?   [250]

Graphs are often used with problems like this. We plot distance moved since the start against the time from the start. The result may be like that shown.

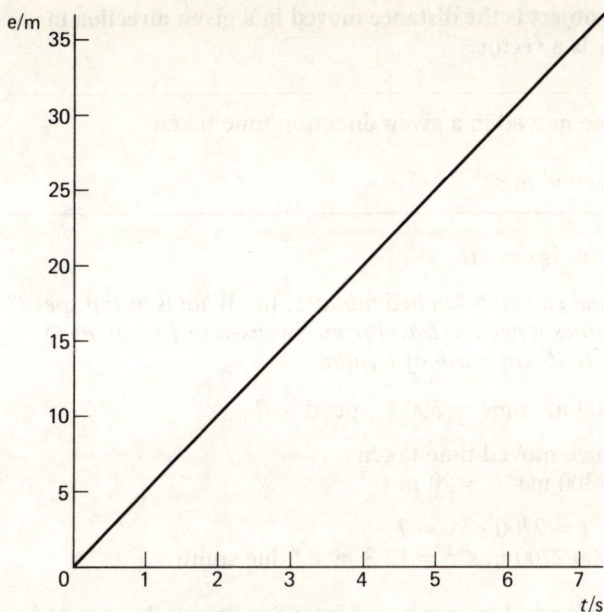

Fig. 1.1 A distance–time graph

The slope of a distance – time graph is the object's speed.

*What is the speed of the object whose motion is plotted in
Fig. 1.1?*   [315]

Of course it is rare for an object's speed to be constant. Only then is
the speed formula given above quite correct. Otherwise it gives **mean
speed**. ('Mean' means 'average'.)

1.2   Speed, velocity and acceleration

As has been said, we are making a study of forces. Forces are **vectors**.
This means that to describe a force fully, we must give its direction as
well as its size. Not all measures are vectors – not all involve direction.
Measures without direction are called **scalars** – mass, wavelength, work
for instance.

A scalar has size but not direction.

A vector has direction as well as size.

Speed is a scalar. When we discuss a car's top speed, we don't mind
which way it is going. But when we talk about forces and motion, we
must use the vector form of speed; this is **velocity**.

The velocity of an object is the distance moved in a given direction in unit time. Velocity is a vector.

Velocity = distance moved in a given direction/time taken
$v = s/t$.
Unit: metre per second m s^{-1}

A useful form of this is $s = vt$.

Flat out, a speedboat can go 6 km in 5 minutes. (a) What is its top speed? (b) What velocity does it need to travel from Swansea to Lynton in 45 minutes? Swansea is 36 km north of Lynton.

(a) distance = 6000 m   time = 300 s   speed = ?

speed = distance moved/time taken
= 6000/300 m s^{-1} = 20 m s^{-1}

(b) s = 36 000 m t = 2700 s v = ?
$v = s/t$ = 36 000/2700 m s^{-1} = 13.3 m s^{-1} due south

Of course this speedboat does not travel from Swansea to Lynton at constant velocity. It starts from rest in Swansea harbour; it comes to rest in Lynton harbour.
    Let us think about the start of that journey. The initial velocity (let us call it v_1) is zero; the final velocity (we'll call this v_2) is 13.33 m s^{-1} due south. There is an **acceleration** as the boat comes to its final velocity. Perhaps the boat takes 20 s to reach this velocity from rest. We can plot these data on a graph (Fig. 1.2).
    Here we assume constant acceleration, and draw a straight-line graph.

Acceleration is the change of velocity in unit time. It is a vector.

Acceleration = change of velocity/time taken
$a = (v_2 - v_1)/t$.   Unit: metre per second per second, m s^{-2}

Acceleration is the slope of a velocity – time graph.

What is the acceleration of the boat whose motion is plotted in Fig. 1.2?

v_2 = 40/3 m s^{-1} due south v_1 = 0 m s^{-1} t = 20 s a = ?
$a = (v_2 - v_1)/t$ = (40/3 – 0)/20 m s^{-2} = 2/3 m s^{-2} due south

A car is moving at 72 km h^{-1} When the brakes are applied it comes to rest in 2 s. What is the acceleration?

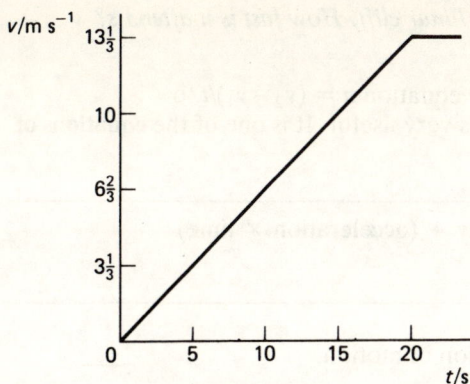

Fig. 1.2 A velocity–time graph

$v_2 = 0$ m s^{-1} $v_1 = 72\,000/3600$ m s^{-1} = 20 m s^{-1} $t = 2$ s $a = ?$
$a = (v_2 - v_1)t = (0-20)/2$ m s^{-2} = $-$ 10 m s^{-2}

A **negative acceleration** means an acceleration in the reverse direction. Thus, if the car is going east when it brakes, the acceleration is westward.

A ball moving at 30 m s^{-1} hit a wall at 90°. It bounced straight back at the same speed. The bounce took 0.2 s. What was the acceleration?

Again we have negative signs with reverse vectors. Here v_2 is in the opposite direction to v_1 – so we must use a negative sign.

$v_1 = 30$ m s^{-1} $v_2 = -30$ m s^{-1} $t = 0.2$ s $a = ?$
$a = (v_2 - v_1)/t = (-30-30)/0.2$ m s^{-2} = -300 m s^{-2}

A girl serves a tennis ball at 10 m s^{-1}. The racquet is in contact with the ball for a quarter second. What is the ball's acceleration?   [200]

We often talk about the acceleration of **free fall**. Any object falling freely (without friction) down to Earth has the same acceleration.

The acceleration of free fall (g) near the Earth's surface is 10 m s^{-2}. You should learn this. The true value is about 9.81 m s^{-2}; 10 m s^{-2} is close in most cases. The size of g in fact depends only on (a) the size of the object whose gravity is acting, and (b) the distance from its surface. Thus on the surface of the Moon (which is much smaller than the Earth) g is about 1.5 m s^{-2}; 6000 km above the Earth g is around 2.5 m s^{-2}.

A stone falls from the top of a cliff. How fast is it moving after 3 s?
[325]

A stone falls from the top of a lunar cliff. How fast is it after 3 s?
[162]

We have just changed the equation $a = (v_2 - v_1)/t$ to
$v_2 = v_1 + at$. This new form is very useful. It is one of the **equations of motion**.

Final velocity = initial velocity + (acceleration × time)
$v_2 = v_1 + at$.

Here is a second useful equation of motion.

Distance moved = mean velocity × time
$s = (v_1 + v_2)/2 \times t$.

A stone falls from the top of a cliff. How far below is it after 3 s?
(g = 10 m s^{-2})　　[240]

A stone falls from the top of a lunar cliff. How far below is it after 3 s?
(g = 1.5 m s^{-2})　　[140]

A car moving at 72 km h^{-1} stops in 2 s. How far does it travel?　　[275]

1.3　Force

Acceleration results if a net force is applied to an object.

What does this mean? It means that if a body's velocity changes, a force must be acting. This is not very precise; we must introduce some more ideas. The first is **mass**.

Mass measures the amount of matter in an object. Mass is a scalar.

Mass relates to **inertia** – how hard it is to accelerate the object. Think of push-starting a large car or a small one. It is harder to accelerate a massive object – such as the large car – than a less massive one. In the same way, it is harder to stop a fat person riding a cycle than a child running at the same speed.

The unit of mass, m, is the **kilogram**, kg.

Do not confuse **mass** and **weight**. (In daily life 'weight' is often used instead of mass. 'I am too fat – I must lose weight.' 'The weight of sugar in this bag is a kilogram.' 'My parcel weighed 2.5 kilograms.') In the free fall of a spacecraft in orbit no weight appears. The craft and its contents are 'weightless'. All the same, objects are still hard to move – they have inertia, or mass.

Weight measures the force of gravity on a mass. Weight is a vector. We return to this later.

The last new measure to discuss is the **momentum** of a moving object. Momentum is rather like inertia; we could say it measures how hard it is to stop the moving object. It depends on the object's mass *and* its velocity.

An object's momentum is the product of its mass and its velocity. It is a vector.

Momentum = mass × velocity

$p = mv$.   Unit: kilogram metre per second, kg m s^{-1}

A 10 t lorry travels at 25 m s^{-1}. What is its momentum?
(1t = 1000 kg)   [222]
A 5 g bee flies at 500 mm s^{-1}. Find its momentum.   [111]

The mass of a brick is 2 kg. It falls for 2 seconds from the top of a building. What is its momentum at the end of that time?   [335]

Now at last we can define **force**. You may have called it 'a push or a pull' before.

Force tends to change an object's momentum. Force is a vector.
The unit of force is the **newton**, N; we define this in the next chapter.

If we reverse the definition of force, we get 'Newton's First Law'. **The momentum of an object is constant unless a net force acts from outside.**
Two problems make it hard to think about this in practice.
The first problem is **friction. Friction is a force that tries to oppose motion**. A car moving on a level road will slow down if the motor stops. The car's momentum gets less – so if Newton's First Law is correct, a force must be acting. That force is friction – from the air, between the wheels and the road, between the wheels and the axles. We often forget friction because we are used to it. Also it cannot be seen.
The second problem is that it is not always easy to detect outside forces. I can make a truck move by pushing from behind. Clearly an outside force pushes the truck. I can also make it move by using a motor. The motor is inside the truck – so how does it act? (If I stand on the truck and push, the truck does not move.)
The motor turns the wheels. The wheels push the ground back – and the ground pushes the wheels forward. The truck has much less mass than the ground – so the truck moves. (Note that the ground will accelerate the other way. But the acceleration is very very small.)

So if the wheels push the ground one way, the ground pushes them the other way. This idea is very useful. It follows from '**Newton's Third Law**'.

If object A applies a force on object B, B applies an equal and opposite force on A.

Let us look at some other cases of force and motion. How do Newton's laws apply?

How does a rocket work? (Fig. 1.3)　　[486]

Fuel and oxygen are pumped into the firing chamber. There they burn to make very hot gases. The rocket motor pushes the gases backward; the gases push the rocket forward. The rocket experiences a net force from outside – it accelerates.
Note: Air is not needed for this. In this case air friction is a nuisance – rockets are more efficient in space.

How does a person take a step? (Fig 1.4)　　[396]

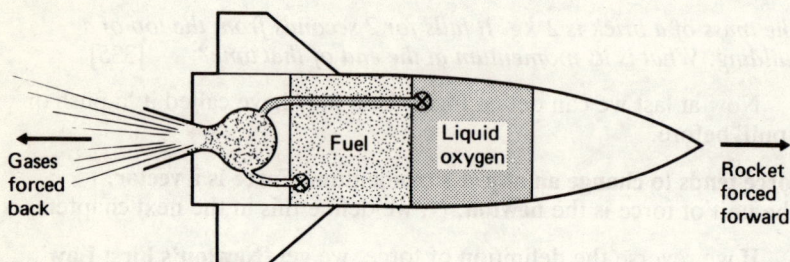

Fig. 1.3 A rocket and Newton's laws

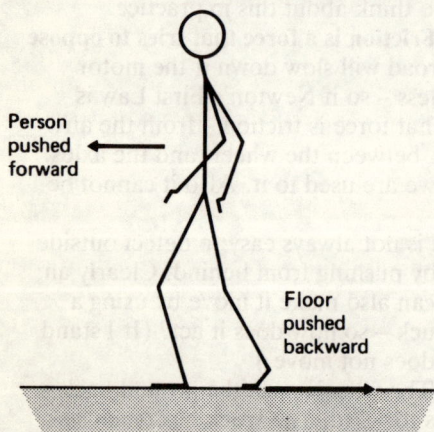

Fig. 1.4 Walking and Newton's laws

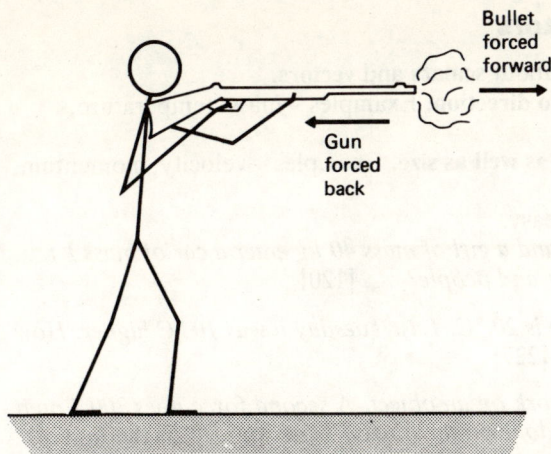

Fig. 1.5 A gun and Newton's laws

Muscles bend the leg forward and raise the heel. The foot pushes the floor backward; the floor pushes the leg – and the person – forward. The person experiences a net force from outside – and accelerates.

How is a bullet fired from a rifle? (Fig. 1.5)   [468]

When the gun fires, it pushes the bullet out. The bullet pushes the gun the other way. (This is called *recoil*.) The gun's backward momentum is taken by the shoulder.

Why can an aircraft not coast at constant speed?   [407]

A plane flying level will slow down if the motors stop. The cause is air friction. The plane pushes the air forward (causing '*wind of passage*'). The air pushes the plane backward. So the plane's forward momentum falls unless the motors are used.

Why does an object in free fall accelerate?   [477]

As the object *does* accelerate, an outside force must be acting on it. This force is the object's **weight** – the '*pull of gravity*' on its mass. If I drop a brick, the brick's weight makes it accelerate to Earth. The Earth pulls the brick downward – so the brick pulls the Earth upward. As a result the Earth accelerates up. However the acceleration is very small indeed and only occurs while the brick is falling.

Think and talk about other cases yourself. How is a moving object made to move? How is it stopped? How do Newton's First and Third laws apply?

1.4 Adding vectors

First remind yourself about **scalars** and **vectors**.
A scalar has size but no direction. Examples – mass, temperature, work.
A vector has direction as well as size. Examples – velocity, momentum, force.

Adding scalars is easy.

A man of mass 60 kg and a girl of mass 40 kg enter a car of mass 1 t. What is the mass of car and people?   [120]

The temperature today is 20 °C. Last Tuesday it was 10 °C higher. How warm was it then?   [227]

One force does 1 kJ work on an object. A second force does 300 J on it. How much work was done on the object? (The joule, J, is the unit of work.)   [327]

Vectors have direction as well as size. Because of this we cannot add them like scalars.

An insect flies at 1 m s⁻¹ in a train moving at 5 m s⁻¹. What is its true velocity?

The answer could be 6 m s⁻¹ if insect and train are moving the same way; 4 m s⁻¹ would be the answer if they move opposite ways. Any other case will give an answer between 4 m s⁻¹ and 6 m s⁻¹.

Two ropes are tied to a pole. Some children pull one with 200 N force; others pull the second with 300 N. What is the total force on the pole?

Again the answer varies – between 100 N (opposite directions) and 500 N (same direction).

So **we cannot add vectors as we add scalars**. We need to know the direction as well as the size of each. Then a scale drawing is used to find the sum. **The sum of vectors is called their resultant**. We can say this another way. **The resultant of vectors is a single vector with the same effect**.

Here is how to find a resultant by scale drawing.

(i) Draw a rough sketch of the vectors in size and direction.
(ii) Choose and record a scale for the main drawing.
(iii) Draw the vectors to scale at the angle given.
(iv) Complete the parallelogram.
(v) Draw the diagonal through the starting point. This relates to the resultant.
(vi) Using the scale and angles in the drawing, measure and describe the resultant.

These steps are used, and marked, in the following questions.

(i) Rough sketch (ii) Scale: 1 cm : 0.5 m s^{-1}
(iii) Scale drawing (iv) Parallelogram (v) Diagonal

5 m s^{-1}

1 m s^{-1}

(vi) *Answer:* The insect moves at 5.1 m s^{-1} at 349° (11° W of N)

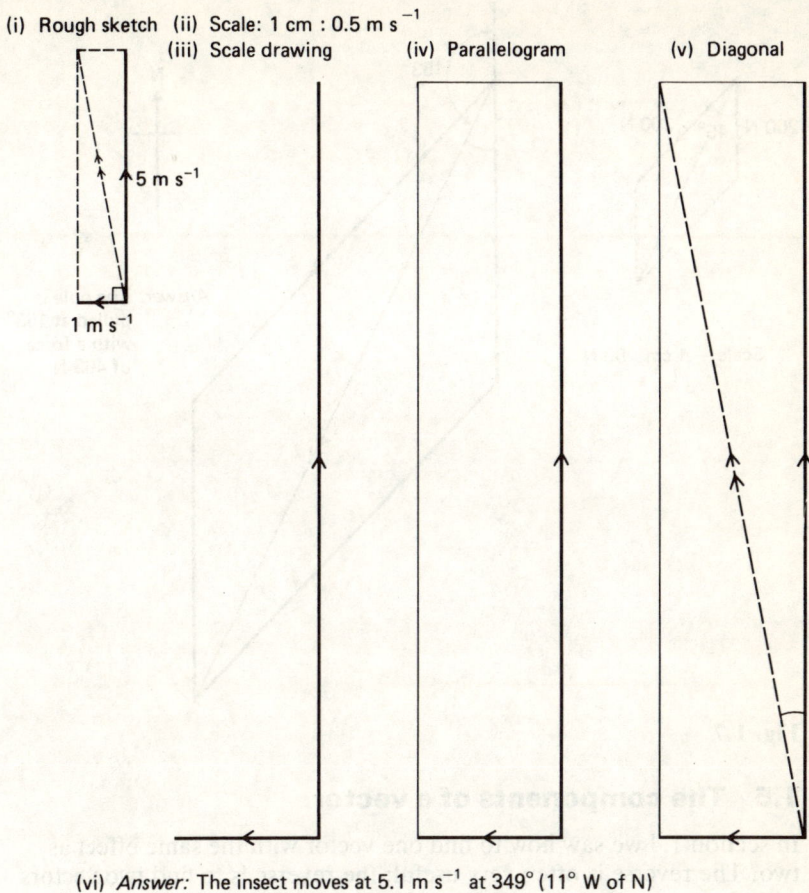

Fig. 1.6 Finding a resultant by scale drawing

An insect flies west at 1 m s^{-1} in a train moving north at 5 m s^{-1}. What is its resultant velocity?   [See Fig. 1.6.]

Two ropes are attached to a pole. Some children pull south on one with a total force of 200 N. Others pull south-east on the second with 300 N. What is the resultant force on the pole?   [See Fig. 1.7.]

Repeat the question about the insect, with it moving in the train (a) northwest [215], (b) southwest [497]. The train still moves north.

Two tugs pull the bow (front) of a ship. One pulls SW with 100 kN; the other pulls SE with 75 kN. What is the resultant force on the bow?   [180]

12

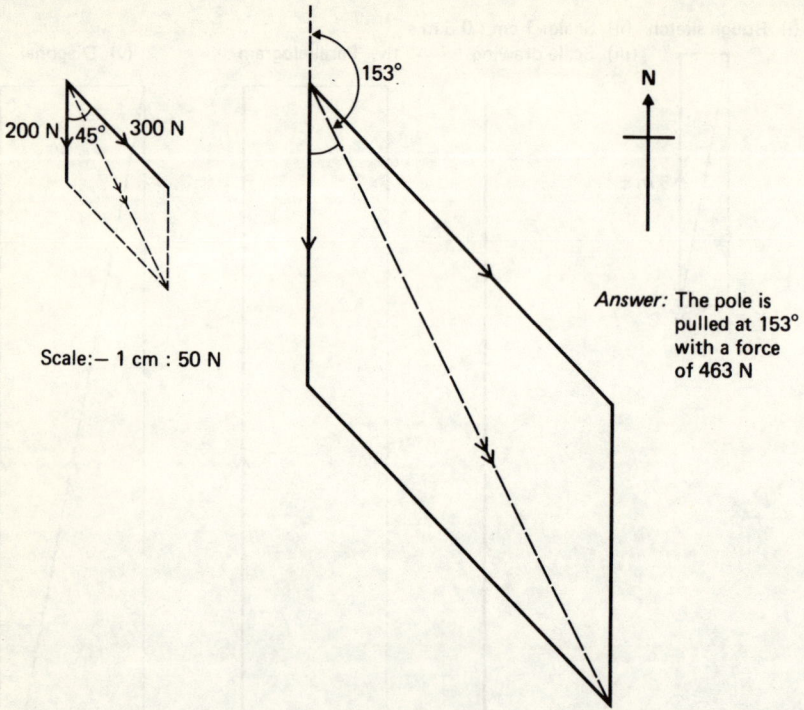

200 N 45° 300 N

153°

N

Scale:— 1 cm : 50 N

Answer: The pole is pulled at 153° with a force of 463 N

Fig. 1.7

1.5 The components of a vector

In section 1.4 we saw how to find one vector with the same effect as two. The reverse is often very useful; the reverse is to find two vectors with the same effect as one. We call this finding the **components** of a vector. Again we must know all the directions.

The components of a vector are the vectors in given directions with the same effect as the one.

In this book we concern ourselves only with components at 90° to each other.

A man pushes one corner of a trolley. The force is 50 N at 40° to the axis of the trolley. What are the forward and sideways components? [See Fig. 1.8.]

We could do this question with a scale drawing based on the rough force diagram (Figure 1.8(b)). However it is simpler and quicker to use trigonometric ratios.

$\sin 40° = F_s/50 \text{ N} \rightarrow F_s = 50 \sin 40°\text{N} = 32.1 \text{ N}$
$\cos 40° = F_f/50 \text{ N} \rightarrow F_f = 50 \cos 40°\text{N} = 38.3 \text{ N}$

(a) Plan view

(b) Force diagram (rough)

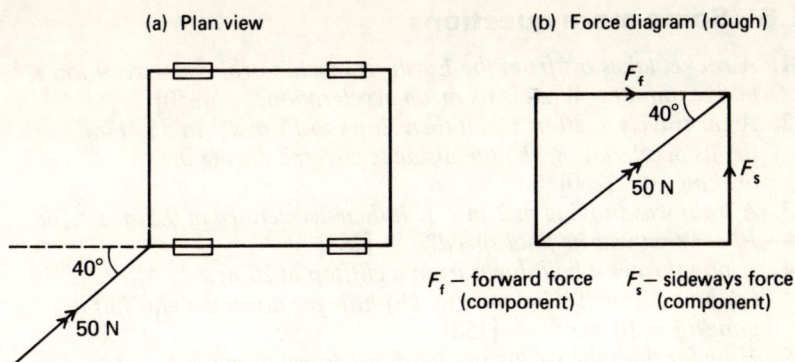

F_f — forward force (component) F_s — sideways force (component)

Fig. 1.8

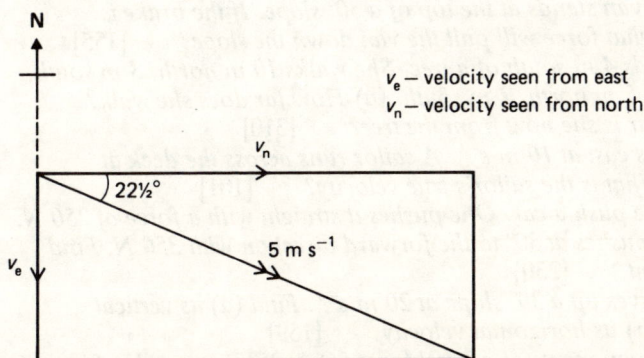

v_e — velocity seen from east
v_n — velocity seen from north

Fig. 1.9

A ship travels at 5 m s^{-1} ESE. How fast will it seem to be going (a) to someone east of it, (b) to someone north of it?   [235]

A cord is tied to the head of an upright nail. The tension in the cord is 200 N; the cord is at 30° to the vertical. What are (a) the horizontal, and (b) the vertical forces on the nail head?   [171]

A shot is fired at 500 m s^{-1} at 25° to the horizontal. What are (a) the vertical, and (b) the horizontal components of its velocity?   [299]

Momentum (p) is a vector too. So an object's momentum can have components. Try this question. We recall that $p = mv$; the unit is kg m s^{-1}.

A 500 g ball moving at 3 m s^{-1} hits a wall at 35°. What are the components of its momentum (a) along the wall, (b) at 90° to the wall?   [135]

1.6 Some more questions

1. *A rocket takes off from the Earth. It reaches orbit velocity, 9 km s⁻¹,* 9 km s^{-1}, *in five minutes. What is its mean acceleration?* [350]
2. *A car travels at 30 m s⁻¹; it then slows to 15 m s⁻¹ in 3 s. Find* 30 m s^{-1} *it then slows to* 15 m s^{-1} *in 3 s. Find* (a) *its acceleration;* (b) *the distance covered during the braking.* [260]
3. *A train was moving at 2 m s⁻¹. It then accelerated at 0.5 m s⁻² for 10 s. What was its final speed?* [285]
4. *A boy throws a ball down from a clifftop at 20 m s⁻¹. After 2 s* (a) *how fast will it be moving,* (b) *how far down the cliff has it gone?* g = 10 m s⁻² [152]
5. *How far does the rocket in question 1 travel to reach orbit?* [308]
6. *A toy balloon is blown up; its neck is not tied. When it is released, it shoots away. Why?* [219]
7. *A 1500 kg van stands at the top of a 30° slope. If the brake is released, what force will pull the van down the slope?* [155]
8. *A girl stands 4 m south of a tree. She walks 10 m north, 3 m south, 2 m north, 5 m north, 2 m south.* (a) *How far does she walk?* (b) *How far is she now from the tree?* [310]
9. *A ship sails east at 10 m s⁻¹. A sailor runs across the deck at 4 m s⁻¹. What is the sailor's true velocity?* [101]
10. *Two people push a car. One pushes it straight with a force of 250 N. The other pushes at 30° to the forward direction with 350 N. Find the resultant.* [230]
11. *A truck drives up a 30° slope at 20 m s⁻¹. Find* (a) *its vertical velocity,* (b) *its horizontal velocity.* [159]
12. *Find the components of a 1 kN force* (a) *at 25° to one side of it,* (b) *at 65° to the other.* [294]
13. *Find the momentum of a 250 000 t ship moving at 20 m s⁻¹.* [117]
14. *The momentum of a 10 t truck is 5 × 10⁵ kg m s⁻¹. What is its velocity?* [147]
15. *Which has greater momentum – a 50 kg man running at 5 m s⁻¹, or a 250 g shot moving at 1 km s⁻¹?* [191]

1.7 Objectives

Note: The first thirteen objectives are from Level 1.
When you have studied this chapter, you should be able to:

(1) define and calculate speed;
(2) plot and use distance–time graphs;
(3) calculate mean speeds given numerical or graphical data;
(4) define acceleration;
(5) plot and use speed–time graphs for motion in a straight line;

(6) describe 'free fall' as constant acceleration under gravity;
(7) solve problems using $s = \frac{1}{2}(v_1 + v_2)t$;
(8) solve problems using $v_2 = v_1 + at$ and $v_2 = v_1 + gt$;
(9) define scalars and vectors, and give examples;
(10) define force;
(11) state that forces, due to gravity, act on masses;
(12) name the unit of force;
(13) determine graphically the resultant of two vectors at a point;
(14) state and explain Newton's First Law;
(15) relate mass to inertia;
(16) distinguish between mass and weight;
(17) define linear momentum;
(18) resolve vectors into two perpendicular directions;
(19) state and explain Newton's Third Law.

Chapter 2

Newton's Second Law

2.1 Newton's three laws

In section 1.3 we met Newton's first and third laws of force.

In chapter 4 we look at both of these in more detail. Now we use the second law – but still keep the other two in mind.

The second law follows on from the first. Here is a new way of thinking about the first law.

If the momentum of an object is constant, the net outside force is zero. If the momentum of an object is not constant, the net outside force is not zero.

On the Earth it is not easy to accept these statements. The forces of **friction** and **gravity** always have an effect. So we cannot really test an object free from forces from outside. Near a spacecraft in free fall, however, neither friction nor gravity have much effect. Next time you see film of 'weightless' objects, check if Newton's First Law is correct.

Now we write the new statements mathematically. We use $\triangle$ (the Greek d, pronounced 'delta') to mean 'change of'. F is force; p is momentum; t is time.

$$\triangle p = 0 \rightarrow F = 0$$

$$\triangle p \neq 0 \rightarrow F \neq 0$$

In fact the force always equals the change of momentum in unit time.

$$F = \Delta p/t$$

Putting this in words gives this statement of **Newton's Second Law**.

An object's change of momentum in unit time equals the net outside force.

In all the work we shall do, an object's mass is constant. In that case, we get a simpler equation as follows.

(a) We have $\qquad\qquad F = \Delta p/t$

(b) So $\qquad\qquad\qquad F = \Delta(mv)/t$, as $p = mv$

(c) As m is constant, $\quad F = m \times \Delta v/t$

(d) Therefore $\qquad\quad F = m\ (v_2 - v_1)/t$ as $\Delta v = v_2 - v_1$

(e) Therefore $\qquad\quad F = ma$, as $a = (v_2 - v_1)/t$

That is a simpler form of Newton's Second Law which covers all normal cases.

We can now write the law like this.

The net outside force equals the product of the object's mass and acceleration.

Force = mass × acceleration
$F = ma$

So here are all three **Newton's laws**.

1. **An object's momentum is constant unless a net force acts from outside.**
2. **The net outside force equals the product of the object's mass and acceleration.**
3. **If object A applies a force on object B, B applies an equal and opposite force on A.**

2.2 The newton

The newton, N, is the SI unit of force. Its definition comes straight from the second law, $F = ma$:

One unit of force = one unit of mass × one unit of acceleration.

or unit force = unit mass × unit acceleration.

So $\qquad$ one newton = one kilogram × one metre per second per second,

or $\qquad\qquad 1\ \text{N} = 1\ \text{kg} \times 1\ \text{m s}^{-2}$

The newton is the force accelerating one kilogram by one metre per second per second.

18

Complete Table 2.1.
Table 2.1.

Force acting	Mass affected	Acceleration produced
(a) 1 N	1 kg	1 m s^{-2}
(b) ? [391]	10 kg	0.1 m s^{-2}
(c) ? [420]	0.5 kg	30 m s^{-2}
(d) ? [329]	1 t (1000 kg)	2 m s^{-2}
(e) ? [181]	100 t	50 m s^{-2}
(f) ? [266]	4 kg	g (10 m s^{-2})
(g) ? [102]	60 kg	g

We have said before that **weight is a force**. Therefore the unit of weight is the newton. The acceleration used to find weight is the acceleration of free fall. This is about 10 m s^{-2} near the Earth's surface.

The weight of an object is ten times its mass. This very useful relation assumes SI units, and applies near the Earth.

Here is the equation for weight.

Weight = mass × acceleration of free fall
$W = mg$.   Unit: newton, N

Because this relation is so useful, g is not just described as acceleration. It is also the ratio of an object's weight to its mass; this is constant in a given place. We call it **gravitational field strength**.

We use the concept of fields quite often in physics, as regions of space in which certain forces may appear. Thus a gravitational force field is a region of space in which there are gravitational forces – a region of space, in other words, in which a mass has weight. No one knows what a gravitational field *really* is, but we believe that they are to be found throughout the universe. Our Sun has a gravitational field; all the objects in the solar system are held by it. The Earth also has a gravitational field which makes apples fall downward and keeps the moon in orbit around it.

The strength of a gravitational field may be measured by the acceleration of masses within it.

We met this idea in section 1.2. On the Moon g is 1.5 m s^{-2}; the lunar gravitational field strength is about 15 per cent of that near the Earth. The Sun's gravitational field strength is roughly 28 times that of the Earth – so near the Sun g would be about 280 m s^{-2}.

So from above we also have

$g = W/m$ Unit: newton per kilogram, N kg^{-1}

g is acceleration of free fall or gravitational field strength. Its unit is m s^{-2} or N kg^{-1}.

The two forms of *g* are not really different; they are different ways of expressing the same thing.

2.3 Using *F* = *ma*

This equation (and its other form *W* = *mg*) is so useful that we need more practice with it. We will also practise use of **SI unit prefixes**. The SI unit prefixes that you should know are in Table 2.2.

Table 2.2 The main SI unit prefixes

Prefix	Symbol	Meaning	Example
pico-	p	10^{-12}, one million millionth	1 pW = 10^{-12} watt
nano-	n	10^{-9}, one thousand millionth	1 ns = 10^{-9} second
micro-	µ	10^{-6}, one millionth	1 µV = 10^{-6} volt
milli-	m	10^{-3}, one thousandth	1 mm = 10^{-3} metre
kilo-	k	10^{3}, one thousand	1 kA = 10^{3} ampere
mega-	M	10^{6}, one million	1 MΩ = 10^{6} ohm
giga-	G	10^{9}, one thousand million	1 GN = 10^{9} newton

Notes:
1. For mass, prefixes attach to -gram not to kilogram. Thus we say one milligram, not one microkilogram. (This is a fault in SI that is still to be settled.)
2. The prefixes centi- (c, 10^{-2}), deci- (d, 10^{-1}) and deka (dk, 10) are quite common in daily life. They are not SI; we should avoid them where possible.

Remember: When you do questions with numbers, the first thing to write is 'given' and 'wanted', *using basic SI units*.

How much would a 10t elephant weigh on the Moon? There g
= 1500 mm s^{-2}.   [432]

A 100 g ball moving at 25 m s^{-1} hit a wall at 90°. It bounced straight back at the same speed. The bounce took 200 ms. What force did the wall apply to the ball?   [351]

A 1 t car moving at 36 km h^{-1} stops in 1 s. What is the braking force?   [449]

The thrust of a 100 t rocket is 2 MN. How long after launch is needed for it to reach 6 km s^{-1}?   [199]

The mass of a belt and gear is 200 kg. It is accelerated from rest to 1.8 km h^{-1} in 0.25 s. Friction is 250 N. What driving force is needed?   [109]

2.4 Some more questions

1. *Relate Newton's three laws. Use as an example driving a car from rest to 50 km h⁻¹.* [440]
2. *A net force of 1 N acts on a 1 kg object for 1 s. What distance does it move?* [201]
3. *Complete the table.*

Table 2.3

Force acting	Mass affected	Acceleration produced
(a) 1 kN	1 kg	1 km s⁻²
(b) ? [182]	1 mg	1 km s⁻²
(c) ? [450]	5 g	*g*
(d) ? [202]	5 g	5 *g*
(e) 50 N	50 g	? [225]
(f) 1 MN	20 t	? [249]
(g) 35 μN	7 mg	? [343]
(h) 20 N	? [409]	5 m s⁻²
(i) 5 kN	? [349]	*g*
(j) 8 pN	? [130]	4 mm s⁻²

4. *A 50 g frog jumps at 2 m s⁻¹. The take off lasts 1 ms. (a) What force do its legs exert on the ground? (b) By how much do its legs stretch?* [359]
5. *The mass of a lift is 1.5 t. What is the tension in the cable when it is accelerating downward at 2.5 m s⁻²? (g = 10 m s⁻²)* [460]
6. *In a TV tube an electron is accelerated from rest to 20 Mm s⁻¹ by 10⁻¹⁴N. The electron mass is 10⁻³⁰ kg. For how long does the force act?* [360] *How far does the particle move?* [494]

2.5 Objectives

Note: The first objective is from Level 1.
When you have studied this chapter, you should be able to

(1) select and use preferred unit prefixes in accordance with SI;
(2) state Newton's Second Law in terms of momentum and as $F = ma$;
(3) describe a gravitational field as a region in which a mass experiences a force;
(4) solve problems using $F = ma$;
(5) define the newton;
(6) determine the weight of a mass, and the mass of an object from its weight;
(7) define and explain g, both as an acceleration and as gravitational field strength.

Chapter 3

Motion in a circle

3.1 Linear and circular motion

Think about the cases shown in Fig. 3.1. In each, an object is moving in a circular path. Up to now we have dealt only with objects moving in a straight line; that is called **linear motion**. Now we look at **circular motion** – movement in a circle or part of a circle.

The figure shows (a) a bicycle wheel, (b) a flywheel (widely used to store energy), (c) a stone in a sling, (d) a spacecraft in orbit round the Earth.
In each case, energy must be used to cause the circular motion. This means that a force is applied and does work.

Now we shall **compare linear and circular motions**. All our ideas about the first can extend to the second. Thus there are special equations of motion and forms of Newton's laws for motion in a circle. Angular velocity, angular acceleration, angular momentum compare to the linear measures; torque compares to force and so on.

Here, however, we shall deal only with circular motion at constant speed, and ignore torque and angular acceleration.

3.2 Angular velocity

A record-player turntable turns at constant speed. *What* is the speed?

A record player table turns 33 times a minute. (a) What is the speed of a point on it 50 mm from the centre? (b) What is the speed of a point on it 150 mm from the centre?

(a)

(b)

(c)

(d)

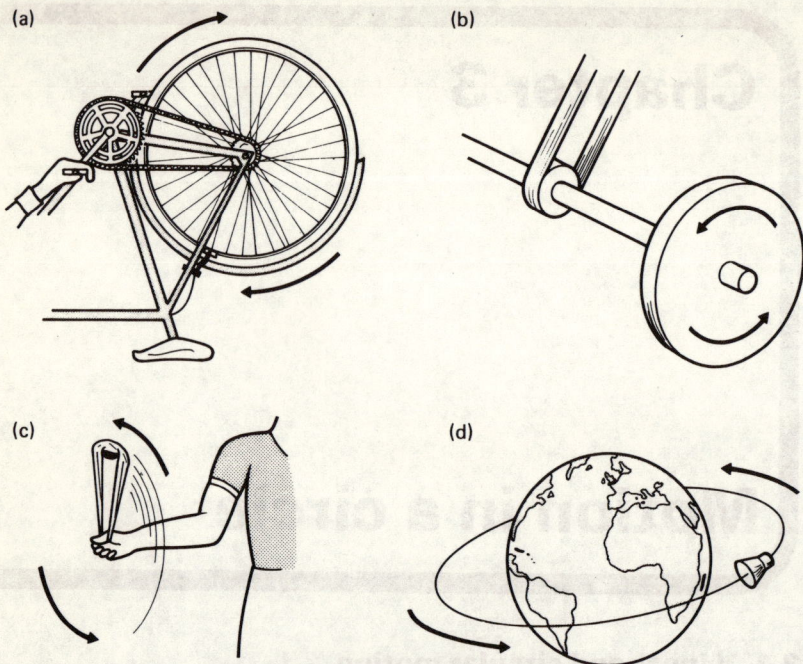

Fig. 3.1 Cases of motion in a circle

(a) Speed = distance moved by point in a second
     = $(33/60) \times 2 \pi \times (50/1000)$ m s^{-1} = 0.17 m s^{-1}
(b) Speed = distance moved by point in a second
     = $(33/60) \times 2 \pi \times (150/1000)$ m s^{-1} = 0.52 m s^{-1}

Thus the concept of speed for motion in a circle cannot be the same as for straight-line motion. What we use instead is **angular velocity**. Figure 3.2 helps to explain this.

When we start timing, let us say the object is at P on its path. As it goes round the circle, the line between it and the centre O moves through an angle. We call this angle θ (the Greek letter th, called theta). If the object moves quickly, θ changes quickly. If it moves slowly the angle changes slowly.

The angular velocity of an object is the change of angle to the centre in unit time.

There are two common **units of angle**. We all know the *degree* (°); this is one ninetieth of a right angle. Here we use the **radian** (rad). The radian is about 57°; there are 2π radians in a circle (360°). If you haven't come across the radian before, Fig. 3.3 should help to explain it. When $s = r$, angle θ is one radian. When $s = 2 \pi r$, angle θ is 2π radian.

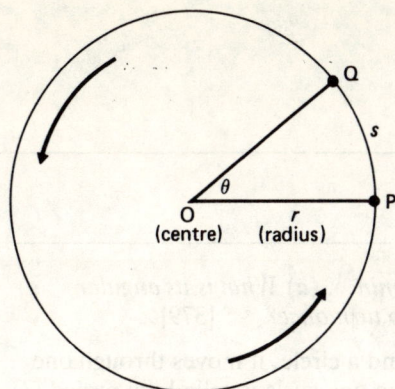

Fig. 3.2 Circular motion

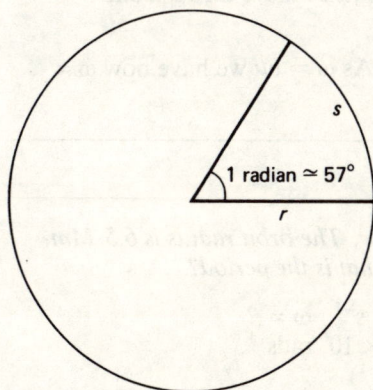

Fig. 3.3 There are 2π radians in 360°

The angle θ, in radians, is defined as the ratio of the arc length s to the radius r.

Angle in radians = arc length radius
$$\theta = s/r$$
Unit: radian, rad

Now   Angular velocity = angle moved/time taken
$$\omega = \theta/t$$
Unit: radian per second, rad s^{-1}

Refer again to Figs. 3.2 and 3.3. We can combine these last two equations.

They give	$\omega = s/r/t$
This leads to	$s/t = r\,\omega$
But	$v = s/t$
Therefore	$v = r\,\omega$

Speed = radius × angular velocity

A record player table turns at 45 rev min^{-1}. (a) What is its angular velocity? (b) How long does it take to turn once? [379]

When an object moves once round a circle, it moves through one **cycle**. The time for the object to make one cycle is called the **period**, T. We define period thus.

The period of a circular motion is the time to move through one complete cycle.

For one cycle, $\theta = 2\pi$ and $t = T$. As $\omega = \theta/t$ we have now $\omega = 2\pi/T$, so $T = 2\pi/\omega$.

$T = 2\pi/\omega$ Unit: second, s

A spacecraft orbits the Earth at 8 km s^{-1}. The orbit radius is 6.5 Mm. (a) What is the angular velocity? (b) What is the period?

(a) $r = 6.5 \times 10^6$ m $v = 8.0 \times 10^3$ m s^{-1} $\omega = ?$
   $v = r\,\omega \rightarrow \omega = v/r = 8 \times 10^3/6.5 \times 10^6$ rads^{-1}
   $= 1.23$ mrads^{-1} (1.23×10^{-3} rads^{-1})
(b) $T = ?$
   $T = 2\pi/\omega = 2\pi/1.23 \times 10^{-3}$ s $= 5100$ s (about 85 minutes)
For the second part, we could have used $s = 2\pi r$ to give the distance round the orbit. If we divide this by the speed, the same value of T results.

A racing car moves at 180 km h^{-1}. The radius of its wheels is 400 mm. What is the angular velocity of each wheel? [279]

3.3 Centripetal force

A force is needed to make an object move in a curve. We return to Newton's first law to deal with this.

An object's momentum is constant unless a net force acts from outside.

If an object moves in a circle at a steady speed, its momentum is *not* constant. This is because momentum is a vector; it involves direction – and this object's direction is always changing.

(a)   (b)   (c)

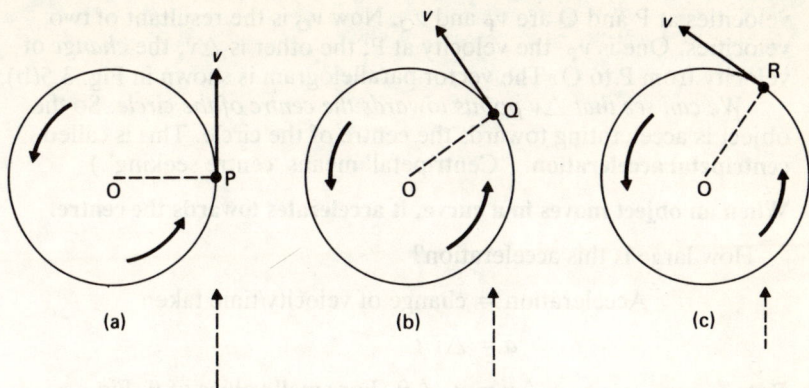

Component of *v* up the page   Component of *v* up the page   Component of *v* up the page

Fig. 3.4 Stages in circular motion

To make an object move in a circle, a net outside force acts.

This idea may not be easy to accept. But it is so important that we must look at it in more detail.

The three pictures in Fig. 3.4 show three stages in an object's motion in a circle. In each case the component of the motion 'up the page' is marked. This is a speed with direction – it is the vector velocity.

So the object's velocity changes, although its speed is constant.

If an object's velocity changes, it is accelerated. **Acceleration** is a vector too. What is the direction of this acceleration?

In Fig. 3.5(a) an object is shown at two *close* points P and Q. The

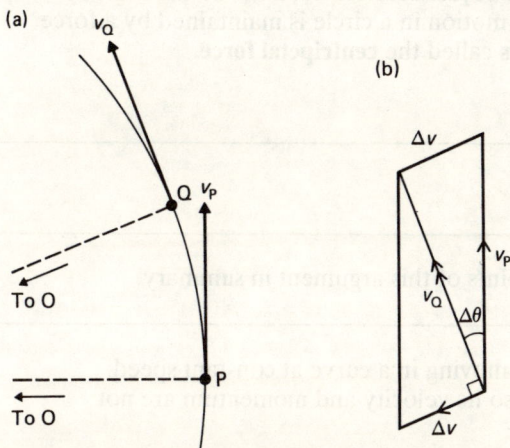

(a)

(b)

Q v_P

To O

To O

P

Δv

v_P

v_Q

$\Delta \theta$

Δv

Fig. 3.5(a) Close stages in circular motion
       (b) The vector diagram

velocities at P and Q are v_P and v_Q. Now v_Q is the resultant of two velocities. One is v_P, the velocity at P; the other is $\triangle v$, the *change* of velocity from P to Q. The vector parallelogram is shown in Fig. 3.5(b).

We can see that $\triangle v$ *points towards the centre of the circle.* So the object is accelerating towards the centre of the circle. This is called **centripetal acceleration**. ('Centripetal' means 'centre seeking'.)

When an object moves in a curve, it accelerates towards the centre.

How large is this acceleration?

Acceleration = change of velocity/time taken

$$a = \triangle v/t$$

But
$$\triangle v = v_P \triangle \theta \text{ (For small values of } \theta, \text{ Fig. 3.5(b), } \tan \theta = \theta)$$

So
$$a = v_P \triangle \theta/t = v \triangle \theta/t, \text{ where } v \text{ is the object's velocity at any instant}$$

Therefore
$$a = v\omega \quad (\text{as } \omega = \triangle\theta/t)$$

or
$$a = v^2/r \quad (\text{as } \omega = v/r).$$

Centripetal acceleration is the square of the velocity divided by the radius of the curve.

Now we can return to forces!

When an object moves in a curve, it accelerates towards the centre; $a = v^2/r$. When an object accelerates, a net outside force acts on it in that direction. So motion in a circle is maintained by a force towards the centre. This is called the **centripetal force**.

$$F = ma$$

Therefore $\quad F = mv^2/r$

Here are the main points of this argument in summary.

1. We think of an object moving in a curve at constant speed.
2. Its direction changes, so its velocity and momentum are not constant.

3. Therefore it is accelerated, towards the centre. This 'centripetal acceleration' is given by $a = v^2/r$. Unit: m/s^2
4. Therefore a force acts on the object, towards the centre. This 'centripetal force' is given by $F = mv^2/r$.

3.4 Centripetal force in practice

In daily life, people often talk of '**centrifugal force**'. Hundreds of years ago, scientists did not understand motion in a circle; centrifugal force is one of their false ideas that linger on.

There is no such thing as centrifugal force!

People say they feel 'centrifugal force' in a car turning a corner. What is really happening?

As a car is moving in a curve, there must be a centripetal force. This is in fact the friction between the road and the tyres. (It is very hard to turn on ice!) The people in the car tend to move on in a straight line. (There's Newton's first law again.) The car moves left; the people go straight – so they *think* they are forced to the right. That is away from the centre. Centrifugal means centre fleeing.

Of course the passengers stay in the car. So they, too, move in the curve. The centripetal force on *them* comes from the car – the seats, the seat belts, the side of the car push them round.

Next time you are driven round a bend, check how you feel about centripetal force.

What provides the centripetal force on a spacecraft in Earth orbit? [289]

What provides the centripetal force on a stone in a sling? [103]

Think in this kind of way about other cases of motion in a curve. What provides the force? Is it always towards the centre? Try to relate each case to the statements at the end of section 3.3. *Never* use the phrase 'centrifugal force'!

A 50 t spacecraft orbits Earth at 8 km s^{-1}. The orbit radius is 6.5 Mm. What is the centripetal force?

$m = 5 \times 10^4$ kg $v = 8 \times 10^3$ m s^{-1} $r = 6.4 \times 10^6$ m $F = ?$

$F = mv^2/r = 5 \times 10^4 \times (8 \times 10^3)^2/6.4 \times 10^6$ N = 500 kN.

Compare this answer with the spacecraft's weight . . . Take g to be 10 m s^{-2}.
Can you explain what you find?

A 1.5 t car turns a corner of radius 100 m at 72 km h⁻¹. Find (a) the centripetal acceleration [251] *(b) the centripetal force.* [160]

A 2 kg stone is whirled in a horizontal circle at the end of a 500 mm string. The string's breaking tension is 400 N. How fast can the stone move without breaking the string?

$m = 2$ kg $r = 0.5$ m $F = 400$ N $v = ?$

$F = mv^2/r \rightarrow$
$v = \sqrt{(Fr/m)} = \sqrt{(400 \times 0.5/2)}$ m s⁻¹ $= 10$ m s⁻¹

*Find the orbital speed for a spacecraft moving round the Moon. Take **g** as 1.5 m/s² and the orbit radius as 1800 km.* [131]

3.5 Some more questions

(Where you need, take *g* as 10 m/s².)

1. *A belt runs on a pulley of radius 450 mm. The pulley turns at 180 rev. min⁻¹. What is the belt speed?* [118]
2. *The Earth spins once in about 86 400 s. Its radius is 6.4 Mm. (a) What is its angular velocity?* [410] *(b) What is the speed of a place on the equator?* [127] *(c) What is the speed of Newcastle (latitude 55 °)?* [429]
3. *The Earth orbits the Sun at 1.5 × 10¹¹ m distance. Its mass is 6 × 10²³ kg. (a) What is the period? (b) What is the angular velocity? (c) What is the speed? (d) What is the centripetal acceleration? (e) What is the centripetal force? (f) What is the acceleration of solar gravity at this distance?* [300]
4. *A 1 kg load on a 1 m string is made to move in a horizontal circle of radius 0.6 m. What are (a) the period, (b) the tension in the string?* [104]
5. *A spaceman in training sits in a box at the end of a 5 m centrifuge arm. He is whirled in a horizontal circle. What is the highest angular velocity allowed if he can stand accelerations up to only 9 g?* [369]

3.6 Objectives

When you have studied this chapter, you should be able to
(1) define angular velocity;
(2) use radian measure;
(3) define the period of a repeated motion;
(4) state the relation between linear and angular velocity;
(5) discuss the need for, and examples of, centripetal force;
(6) derive $F = mv^2/r$ and solve simple problems involving it.

Chapter 4

Momentum and kinetic energy

4.1 Linear momentum again

The concept of momentum was met and defined in Chapter 1; it has been used often since. Now let us look at it in more detail. Throughout we shall be concerned with linear momentum – i.e. the momentum of an object moving in a straight line.

We said that momentum measures how hard it is to stop a moving object. We also defined it.

An object's momentum is the product of its mass and its velocity. Momentum is a vector.

Momentum = mass × velocity
$p = mv$ Unit: kilogram metre per second, kg m s^{-1}

A 50 g bullet is fired at 1 km s^{-1}. Find its momentum.   [150]

A 40 kg girl walks at 3.6 km h^{-1}. What is her momentum?   [218]

It can be harder to stop a bullet than a walking person!

A 100 t spacecraft moves at 8 km s^{-1}. Find its momentum.   [370]

4.2 Conservation of momentum

From Newton's First and Third Laws we derive the important **law of constant momentum**. Sometimes this is called the *principle of conservation of momentum*.

The total momentum of a group of objects is constant, unless a net outside force acts.

Clearly this is a second form of Newton's First Law. It applies to a group of objects rather than to one.

'A group of objects' means a set of two or more things connected in some way. The rifle and the bullet are a group of objects. So is the girl and the ground. So is the rocket and its exhaust.

We must also note that momentum is a vector. This means that to find total momentum, the directions of the components must be taken into account.

Follow this example with care.

A 10 kg rifle fires a 50 g bullet at 1 km s^{-1}. With what velocity does the rifle recoil?

(i) The group of objects is the gun and the bullet.
(ii) Before firing, both were at rest. Total momentum was zero.
(iii) After firing, the bullet goes one way with a certain momentum.
(iv) The total momentum of the group doesn't change; the total momentum remains zero.
(v) So the gun acquires momentum the other way.

$m_r = 10$ kg $\quad V_{1r} = 0$ m s^{-1} $\quad m_b = 0.05$ kg $\quad v_{1b} = 0$ m s^{-1}
$v_{2b} = 1000$ m s^{-1} $\quad v_{2r} = ?$

Subscript$_r$ refers to the rifle; $_b$ is for the bullet.
Momentum before firing = momentum after firing.

$$m_r v_{1r} + m_b v_{1b} = m_r v_{2r} + m_b v_{2b}$$

$$(10 \times 0) + (0.05 \times 0) = 10 v_{2r} + (0.05 \times 1000) \text{ m s}^{-1}$$

$$\rightarrow v_{2r} = -0.05 \times 1000/10 \text{ m s}^{-1} = -5 \text{ m s}^{-1}$$

Thus the rifle recoils at 5 m s^{-1}.

It is worth checking the above result against Newton's Third Law. The Second Law comes in here. We use the proper form rather than the simple one. (See section 2.1.)

An object's change of momentum in unit time equals the net outside force.

$$F = \Delta p_r / t$$

For the bullet: $\Delta p_b = m_b v_{2b} - m_b v_{1b} = m_b (v_{2b} - v_{1b})$
$= 0.05 \times 1000$ kg m s^{-1} = 50 kg m s^{-1}

For the rifle: $\Delta p_r = m_r\, v_{2r} - m\, v_{1r} = m_r\,(v_{2r} - v_{1r})$
$\qquad\qquad\quad = 10 \times (-5)\ \text{kg m s}^{-1} = -50\ \text{kg m s}^{-1}$

As we expect from the **law of constant momentum**, the momenta are equal and opposite. (*Note*: The plural of momentum is momenta.)

In the above force equation, t is the time during which the force acts. The time for the rifle to push the bullet equals that for the bullet to push on the rifle. Let us say that that time is 0.1 s.

For the bullet: $F_b = \Delta p_b/t_b = 50/0.1\ \text{N} = 500\ \text{N}$

For the rifle: $F_r = \Delta p_r/t_r\ = 50/0.1\ \text{N} = 500\ \text{N}$

We would obtain the same kind of result whatever value of t was chosen. We always find that the two forces are equal and opposite. This is Newton's Third Law.

If object A applies a force on object B, B applies an equal and opposite force on A.

A 20 kg child standing on a 2 kg skateboard jumps off at 1 m s^{-1}. How fast does the skateboard recoil?

$m_c = 20\ \text{kg}\quad m_b = 2\ \text{kg}\quad v_{1c} = v_{1b} = 0\ \text{m s}^{-1}\quad v_{2c} = 1\ \text{m s}^{-1}$
$v_{2b} = ?$

Momentum before jump = momentum after jump

$m_c\, v_{1c} + m_b\, v_{1b} = m_c\, v_{2c} + m_b\, v_{2b}$

$\rightarrow v_{2b} = (m_c\, v_{1c} + m_b\, v_{1b} - m_c\, v_{2c})/m_b$
$\qquad = (20 \times 0 + 2 \times 0 - 20 \times 1)/2\ \text{m s}^{-1} = -10\ \text{m s}^{-1}$

(The board recoils in the other direction.)

A 60 kg man jumps from a 300 kg boat at 2½ m s^{-1}. (a) What is the boat's change of momentum? (b) How fast does it recoil?   [411]

There are various tests of the law of constant momentum we can try in the laboratory. In each case we need to know how a system of objects interact. The momentum of each object, before and after interaction can be obtained by measuring the appropriate mass and velocity.

The masses are determined by suitably accurate weighing. The velocities require to be measured accurately over fairly short distances. This can be achieved in the following way. A thin piece of card is mounted on each object so that the card interrupts a beam of light between a source and a photocell. As the front of the card passes the light the beam is cut off, but reappears as the end of the card goes through. An electronic clock can be used to measure the change in output as the card enters and leaves the beam. The velocity of the object is obtained by dividing the length of the card by the time it takes to pass through the beam.

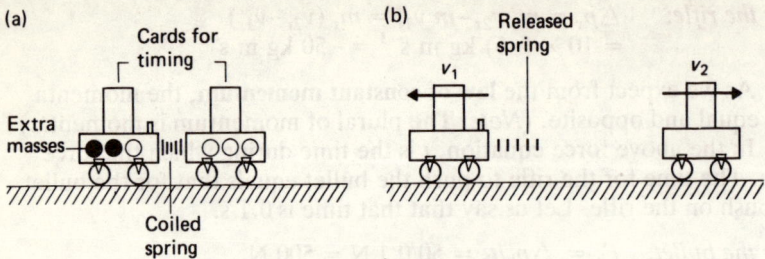

(a)

Cards for timing

Extra masses

Coiled spring

(b)

Released spring

v_1

v_2

Fig. 4.1

'Trolleys' may be used in this way (see Fig. 4.1). Their masses can be altered by adding or removing the metal loads. As shown in Fig. 4.1(a) two trolleys, one carrying a compressed spring, touch each other and are at rest on a level surface – the total momentum is zero. When the spring is released the first trolley pushes the second away but it also recoils (Fig. 4.16). The resulting velocities are determined and the value of

$m_1, v_1 + m_2, v_2$ ($m_1 m_2$ are the masses and $v_1 v_2$ the velocities of the first and second trolley respectively).

calculated. This should be zero if the law of constant momentum is true.

In the next section we will apply the law to cases in which two or more objects collide. We can use trolleys to test collisions but we would then have to measure their velocities before as well as after interaction.

4.3 Collisions

The cases we have looked at so far have all involved separation. The **law of constant momentum** also applies to collisions. We use it in just the same way.

A 50 g bullet moving at 1 km s^{-1} hits a 2 kg skateboard coming towards it at 10 m s^{-1}. The bullet sticks in the skateboard. Describe the motion after impact. [337]

$m_b = 0.05$ kg $v_{1b} = 1000$ m s^{-1} $m_s = 2$ kg
$v_{1s} = -10$ m s^{-1} $v_{2b} = v_{2s} = v_2 = v = ?$

momentum before impact = momentum after impact

$m_b v_{1b} + m_s v_{1s} = m_b v_{2b} + m_s v_{2s}$
$\rightarrow v_2 = (m_b v_{1b} + m_s v_{1s})/(m_b + m_s)$
$= (0.05 \times 1000 \times 2 \times (-10))/2.05$ m s^{-1} = 14.6 m s^{-1}.

After impact the board retreats about 15 m s^{-1}.

Repeat the above question. (a) The skateboard is at rest before impact. [247] *(b) The board and bullet are moving the same way before impact.* [298]

A 1 t car moving north at 108 km h^{-1} collides head-on with a 10 t truck coming south at 27 km h^{-1}. What happens to the combined wreckage?

$m_c = 1000$ kg $v_{1c} = 30$ m s^{-1} $m_t = 10\,000$ kg $v_{1t} = -7.5$ m s^{-1}

$v_{2c} = v_{2t} = v_2 = ?$

momentum before impact = momentum after impact

$m_c v_{1c} + m_c v_{1t} = m_c v_{2c} + m_1 v_{2t}$

$\rightarrow v_2 = (m_c v_{1c} + m_1 v_{1t})/(m_c + m_1) =$
$(1000 \times 30 - 10\,000 \times 7.5)/11\,000$ m s^{-1} = -4.1 m s^{-1}

Repeat the above question. This time assumes (a) the car is at rest before impact, [333]; *(b) the lorry is at rest before impact* [337]; *(c) the car and the lorry are travelling the same way at the velocities quoted.* [389] *Find the car's change of momentum in each case.*

4.4 Kinetic energy

A second major measure of motion is **kinetic energy**. It relates closely to momentum in effect; but the two are *not* the same.

How do they differ? We start with some familiar concepts.

A force does work if it causes movement.

The work done by a force is the product of the force and the distance moved.

Work = force applied × distance moved
$W = Fs$ Unit: joule, J

The joule is the work done by one newton moving one metre.

A man pushes a trolley with a force of 50 N; he moves it 50 m. What work does he do? [301]

A 10 kg block falls from the top of a 35 m building. What work does its weight do? [400]

A space rocket's engines produce 1 MN thrust. They move the spacecraft 100 km. What work is done? [287]

An object's energy is the work it can do.

'Energy is work.' So they have the same symbol, W; they have the same unit, the joule, J.

There are many kinds of energy. Here are some.

Light carries electromagnetic energy. Plants use this to produce food. Sunlight pushes matter from a comet's 'head' to form the 'tail'. Photocells make meter needles move.

Food has chemical energy. Without food we could not do work.

Fuels have chemical energy as well. The energy from fuels does most of the work in our lives.

Sound is mechanical energy. It moves our eardrums; it can break windows.

A wound ("strained") clock spring has potential energy. This makes the clock work for some time. An object at a height has potential energy too – if let fall, as in a piledriver, it can do work. "Potential" means, in effect, "stored".

An object has kinetic energy because of its motion. A moving car can break a wall. A moving stone can break a window. Moving windmill sails can grind corn or drive machines.

To get an object at rest moving at velocity v, a force F is needed. While it acts, force and object move a distance s. The object's kinetic energy will be the work done (W) in accelerating it. Here m is the object's mass.

	$W = Fs$	
So	$W = mas$	(as $F = ma$)
But	$s = \frac{1}{2}(v_1 + v_2)\,t$	(section 1.2)
So	$W = \frac{1}{2}\,mat\,(v_1 + v_2)$	
But	$at = v_2 - v_1$	(from $v_2 = v_1 + at$)
So	$W = \frac{1}{2}m\,(v_1 + v_2)(v_2 - v_1)$	
But here	$v_1 = 0$	
So	$W = \frac{1}{2}mv_2^2$	

An object's kinetic energy is half the product of its mass and the square of its velocity.

Kinetic energy = ½ × mass × velocity²

$W = \frac{1}{2}mv^2$
                                                        Unit: joule, J

A 50 g rifle bullet is fired at 1 km s⁻¹. What is its kinetic energy?

$m = 0.05$ kg $v = 100$ m s⁻¹ $W = ?$

$W = \frac{1}{2}\,m\,v^2 = \frac{1}{2} \times 0.05 \times 1000^2$ J $= 25$ kJ

A 40 kg girl walks at 3.6 km h⁻¹. Find her kinetic energy. [311]

A 100 t spacecraft moves at 8 km s⁻¹. What is its kinetic energy? [184]

4.5 The conservation of energy

There is a law of constant energy, rather as there is with momentum. But it is not a law of constant *kinetic* energy.

This question shows that kinetic energy is not conserved.

A 10 kg rifle fires a 50 g bullet at 1 km s⁻¹. Find the total kinetic energy before and after the shot.

Before: $m_r = 10$ kg $v_{1r} = 0$ m s⁻¹ $m_b = 0.05$ kg $v_{1b} = 0$ m s⁻¹
$W_1 = ?$
$W_1 = \frac{1}{2} m_r v_{1r}^2 + \frac{1}{2} m_b v_{1b}^2 = 0$ J

After: $m_r = 10$ kg $v_{2r} = -5$ m s⁻¹ $m_b = 0.05$ kg
$v_{2b} = 1000$ m s⁻¹ (found in section 4.2.) $W_2 = ?$

$W_2 = \frac{1}{2}m_r v_{2r}^2 + \frac{1}{2}m_b v_{2b}^2$
$= \frac{1}{2} \times 10 \times -5^2 + \frac{1}{2} \times 0.05 \times 1000^2$ J
$= 25$ kJ

Note: **Energy is a scalar**: direction is not involved.

So kinetic energy is *not* constant in a case like this. The kinetic energy that appears during firing comes from burning the fuel (cordite) – an equal amount of energy stored in the fuel is released as it burns.

The **law of constant energy** is as follows. (Sometimes it is called the *principle of conservation of energy*.)

The total energy of an object or group of objects is constant, unless energy is transferred to or from outside.

The following form may be easier to learn.

Energy cannot be created or destroyed.

As far as we know, this is true – as long as all kinds of energy are included.

When two objects collide, for instance, it is not normal for the total kinetic energy to remain constant. Some of the initial kinetic energy will go to raise the temperature. (Think how hot a hammer head becomes in use.) Some part will make sound. Perhaps work will be done to change the shape of one object or the other. What the law tells us is that the *total* energy, counting all forms, does not change.

In a few cases, kinetic energy does stay constant in an *impact*. These are called **elastic collisions**. Collisions between some simple subatomic particles (such as protons or electrons) are thought to be elastic. Collisions between steel balls are almost elastic.

Objects hitting each other when kinetic energy is not conserved have **inelastic collisions**. If the objects stick together after impact, the collision is totally inelastic. This may happen, for instance, with two lumps of putty.

But linear momentum is constant in *all* collisions.

4.6 Some more questions

1. *A crane lifts a 5 t load at 3 m s^{-1}. (a) Find the momentum. (b) Find the kinetic energy.* [106]
2. *A 5 kg mass is at rest. A net 20 N force acts on it for 78 m. (a) What is the acceleration? (b) What is the final kinetic energy? (c) What is the final velocity?* [217]
3. *This object now collides with a 10 kg mass. This was moving the same way at 3 m s^{-1}. The two objects stick to each other. (a) What is their final velocity? (b) What is the change in momentum of each object? (c) What kinetic energy is 'lost' in the impact; what happens to it?* [234]
4. *A 1.5 t car travels at 10 m s^{-1}, then accelerates steadily at 1.0 m s^{-2} to a speed of 15 m s^{-1}. Find (a) the distance covered during the acceleration, (b) the kinetic energy 100 m after the start of the acceleration.* [399]
5. *A 4.0 kg ball moving at 10 m s^{-1} collides with and sticks to a 10 kg ball moving at 4.0 m s^{-1} (a) the same way, (b) the other way. In each case find the velocity after impact and the energy 'loss' during impact. Is energy really 'lost'?* [110]
6. *A spacecraft of mass 200 kg moving at 5.0 km s^{-1} fires a 10 kg probe forward at 1.0 km s^{-1}. What will be the new velocity of the spacecraft?* [261]

4.7 Objectives

Note: The first three objectives are from Level 1.
When you have studied this chapter, you should be able to
  (1) define and calculate work in terms of force applied and distance moved;
  (2) define the joule;
  (3) outline the relation between different forms of energy and work;
  (4) state the law of constant momentum (conservation of momentum);
  (5) solve problems involving change and conservation of momentum;
  (6) describe kinetic energy as the energy due to an object's motion;
  (7) derive and use $W = \frac{1}{2} mv^2$;
  (8) state the law of constant energy (conservation of energy);
  (9) distinguish between elastic and inelastic collisions.
(10) describe tests of the law of constant momentum.

Chapter 5

Work and energy

5.1 A graphical approach to work

The concept of work was revised in section 4.4. Here are the main points again.

A force does work if it causes movement.

The work done by a force is the product of the force and the distance moved.

The joule is the work done by one newton moving one metre.

Here is a **force–distance graph** (Fig. 5.1). It shows that a 5 N force moved an object by 8 m.

Use $W = Fs$ *to find the work done by the force.*   [107]

This answer is the same as the shaded area in the graph. We can expect this. The area of a rectangle is the product of the sides; the sides of the shaded rectangle are 5 N force and 8 m distance.

Area = length × width (Fs) = 5 × 8 Nm = 40 J

The work done by a force is the area between the force–distance graph and the distance axis.

The force–distance graph in Fig. 5.1 is a straight line parallel to the s-axis. That denotes a constant force. In practice there are many cases where the force applied is not constant. Then the graph will not

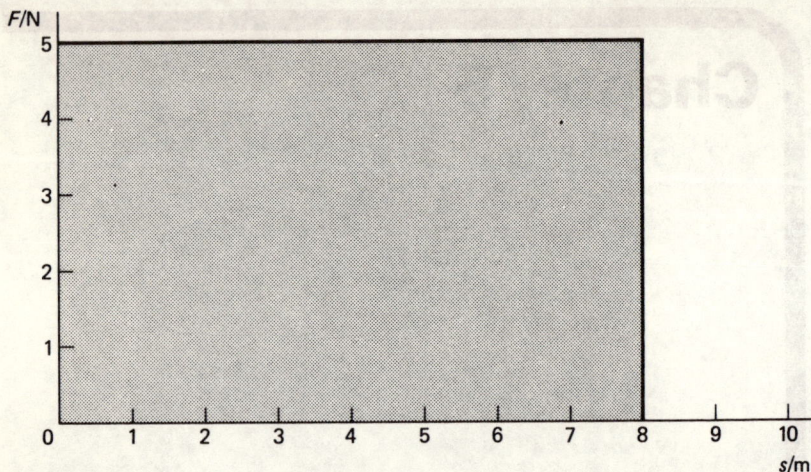

Fig. 5.1 A force–distance graph

be like that shown. All the same, the area between the force–distance curve and the *s*-axis always gives the work done.

Figure 5.2 shows a spring 'chest expander'. The tension in the spring increases by 1 N for each 10 mm it is extended. How much work is done if it is stretched by 500 mm?

We can draw a force–distance graph of this. (See Fig. 5.3.)

The work the girl does is the shaded area. Recall that the area of a triangle is ½ × base × height. Here this gives us 12.5 J.

Of course we could also do this question using $W = Fs$. There is a little catch – the force is not constant this time. So F must be the *mean* force, 25 N.

$F = 25$ N $s = 0.5$ m $W = ?$

$W = Fs = 25 \times 0.5$ J $= 12.5$ J

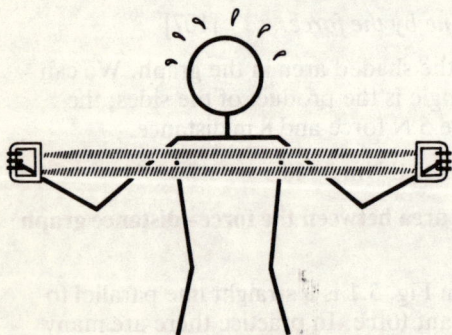

Fig. 5.2 Using a chest expander

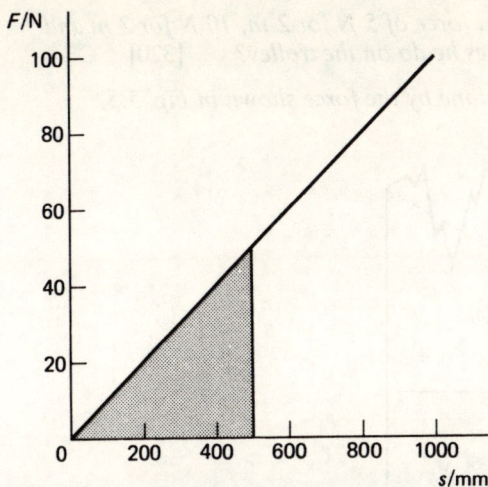

Fig. 5.3 How the chest expander works

Figure 5.4 shows how the force applied to a train by the engine varies with distance from a station. What work is done in the first kilometre?

Again, the work done is the area between the force–distance graph and the distance axis. This is the sum of the areas shown in the figure – areas (a), (b), (c), (d), (e) – total 7.2 MJ.

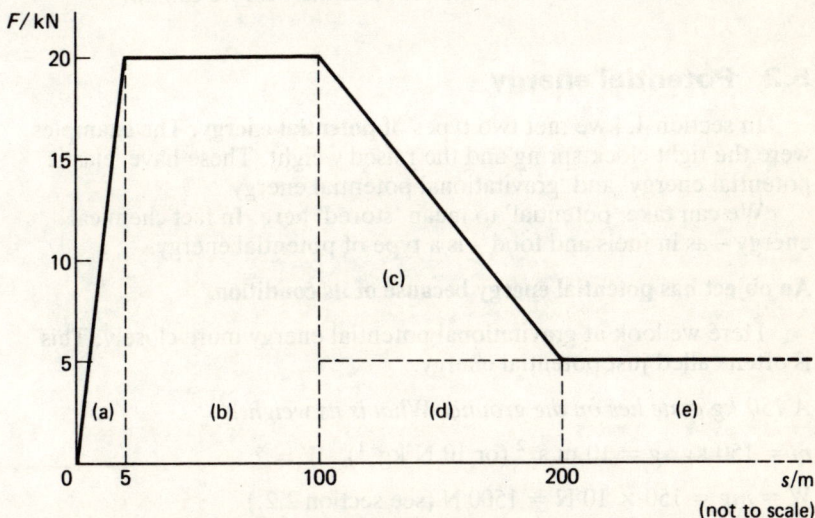

Fig. 5.4 The start of a train journey

40

A man pushes a trolley with a force of 5 N for 2 m, 10 N for 2 m and 20 N for 5 m. What work does he do on the trolley? [320]

Roughly how much work is done by the force shown in Fig. 5.5?

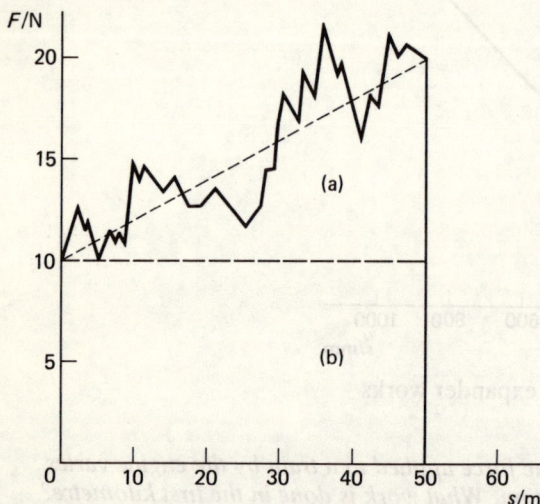

Fig. 5.5 A very variable force–distance curve

750 J. This is done by simplifying the curve to that shown dotted in Fig. 5.5. In that way we may come fairly close to drawing the average situation. The work done is found by adding areas (a) and (b). Such a method will not be fully accurate, but it is the best we can do.

5.2 Potential energy

In section 4.4 we met two types of **potential energy**. The examples were the tight clock-spring and the raised weight. These have 'elastic potential energy' and 'gravitational potential energy'.

We can take 'potential' to mean 'stored' here. In fact chemical energy – as in fuels and food – is a type of potential energy.

An object has potential energy because of its condition.

Here we look at gravitational potential energy more closely. This is often called just potential energy.

A 150 kg crate lies on the ground. What is its weight?

$m = 150$ kg $g = 10$ m s^{-2} (or 10 N kg^{-1}) $W = ?$

$W = mg = 150 \times 10$ N $= 1500$ N (see section 2.2.)

Note: We do not have enough letters for all the symbols needed. W stands for weight as well as for work or energy.

A crane lifts the crate 20 m above the ground. How much work is done?

$m = 150$ kg $a = 10$ m s^{-2} $s = 20$ m $W = ?$

$W = Fs = mgs = 150 \times 10 \times 20$ J $= 30$ kJ

To lift the crate, the crane's motor uses fuel. The chemical potential energy released does work on the crate. So the crate gains energy. The energy gained equals the work done in lifting it. That energy is the crate's new potential energy.

If the crane cable breaks, the load falls. As it falls, it loses potential energy. However, it accelerates – so its kinetic energy increases. When it hits the ground, it does work – it causes damage.

In this example the object's potential energy equals the work done in raising it. We can see that the potential energy would be twice as much if the object had twice the mass, or if it had been raised twice as high, or indeed if g (the gravitational field strength) were twice as big.

These ideas lead to the equation for gravitational potential energy.

Potential energy = mass × acceleration of free fall × height

$W = mgh$ Unit: joule, J

Now mg is the object's weight (a force): h is the distance it would fall. So the energy it has is the work it could do in falling – force × distance, mgh.
From this we can get a definition.

The gravitational potential energy of an object is the product of its weight and its height.

A 100 t spacecraft orbits 100 km up. What is its potential energy?   [241]

A 2 kg cat sits on a tree branch 3 m above ground. Find its potential energy.   [108]

A 20 kg block falls, and hits the ground at 40 ms^{-1}. From what height did it fall?   [151]

5.3   Energy conversion

The idea of energy conversion has become clearer in the last few sections. The last question could be solved using it like this.

$m = 20$ kg $v_2 = 40$ m s^{-1} $h = ?$ $g = 10$ m s^{-2} $v_1 = 0$ m s^{-1}

$mgh = \tfrac{1}{2} mv_2^2$

$\rightarrow h = v_2^2/2g = 40^2/2 \times 10$ m $= 80$ m

This assumes (as does any approach) that no energy is lost by friction.

If this is so, the block's initial potential energy (*mgh*) *is wholly changed to kinetic energy* ($\tfrac{1}{2} mv_2^2$) during the fall.

The idea of energy conversion relates to the **law of constant energy**; we met this in section 4.5.

The total energy of an object or a group of objects is constant, unless energy is transferred to or from outside.

Recall the crane and the crate.
We started with chemical potential energy in the fuel.

(i) As the fuel burned in the motor, energy was released to the crane drive. The crane did work on the crate; this was raised – it gained potential energy.

(ii) The cable broke – the crate's potential energy was changed to kinetic energy as it fell.

(iii) When the crate hit the ground, work was done in causing damage. The simple energy flow diagram in Fig. 5.6 shows these changes.

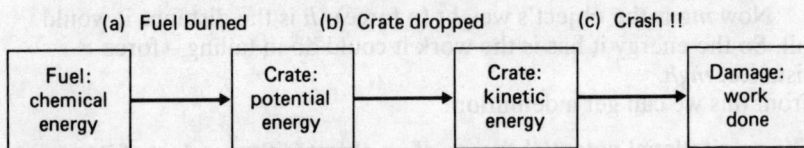

(a) Fuel burned   (b) Crate dropped   (c) Crash !!

| Fuel: chemical energy | → | Crate: potential energy | → | Crate: kinetic energy | → | Damage: work done |

Fig. 5.6 A simple energy flow diagram

If all the changes were perfect, we could have this:

(i) Fuel burned   30 kJ released.
    Crate lifted   30 kJ potential-energy gained.
(ii) Crate dropped   30 kJ kinetic energy.
(iii) Crash!!   30 kJ damage done.

But energy transfers are rarely perfect. Energy is 'lost' on the way. The flow diagram in practice would be more like Fig. 5.7.

Less than a quarter of the energy in the fuel goes to break up the crate.

Of course damage is not a normal aim of fuel use. But the idea of **imperfect energy transfer** is very important. It relates to the **efficiency** of a system.

The efficiency of a system is the ratio of the useful energy output to the energy input.

(a) Fuel burned   (b) Crate dropped   (c) Crash !!

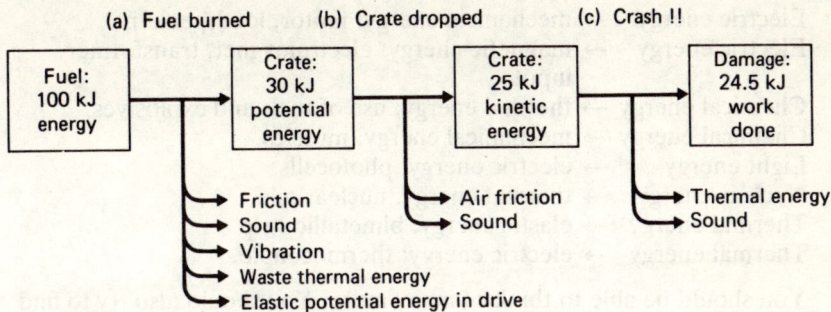

Fig. 5.7 A more likely energy flow diagram

Efficiency = energy output/energy input
$\eta = W_2/W_1$

The symbol for efficiency is η, the Greek e, pronounced 'eta'. There is no unit – the measure is a ratio of energies. However, the fraction is often given as a percentage.

The 'monkey' of a piledriver has a mass of 100 kg. It is raised 4 m by the burning of 0.2 g of oil. The energy value of the fuel is 45 MJ kg^{-1}. When released, the monkey drives the pile 250 mm against a mean friction of 12 kN. (a) Find the efficiency of raising the monkey. (b) What is the piledriving efficiency? (c) What is the total efficiency?

(The symbols used are standard: u_f is the 'energy content' of a kilogram of the fuel. Subscripts $_{1,2}$ and $_3$ refer to the first, second and third energy changes.)

(a) $m_f = 2 \times 10^{-4}$ kg $u_f = 45 \times 10^6$ J kg^{-1}
   m_m 100 kg $g = 10$ m s^{-2} $h = 4$ m $_1\eta_2 = ?$
   $_1\eta_2 = W_2/W_1 = m_m gh/m_f u_f = 100 \times 10 \times 4/(2 \times 10^{-4} \times 45 \times 10^6)$
   $= 0.44$ (or 44 per cent)

(b) $W_2 = 100 \times 10 \times 4$ J
   $F = 12 \times 10^3$ N $s = 0.25$ m $_2\eta_3 = ?$
   $_2\eta_3 = W_3/W_2 = Fs/W_2 = 12 \times 10^3 \times 0.25/(100 \times 10 \times 4) = 0.75$
   (or 75 per cent)

(c) $_1\eta_3 = ?$
   $_1\eta_3 = W_3/W_1 = {_1\eta_2} \times {_2\eta_3} = 0.4 \times 0.75 = 0.3$ (or 30 per cent)

There are very many cases of energy transfer in daily life. Here are some.

Electric energy → thermal energy: fuse, electric fire, filament lamp.
Electric energy → light energy: filament lamp, laser.

Electric energy → mechanical energy: motor, loudspeaker.
Electric energy → magnetic energy: electromagnet, transformer input.
Chemical energy → thermal energy: use of fuels and explosives.
Chemical energy → mechanical energy: muscles.
Light energy → electric energy: photocells.
Nuclear energy → thermal energy: nuclear pile.
Thermal energy → elastic energy: bimetallic strip.
Thermal energy → electric energy: thermocouple.

You should be able to think of many more. You should also try to find out some conversion efficiencies in theory and in practice.

We cannot solve problems about energy changes unless we know the efficiency of the required conversion. Often we assume it is 100 per cent.

Let us look at more questions. First recall some other basic ideas.

Power is the energy transferred or work done in unit time.

Power = energy transferred/time taken
$P = W/t$ Unit: watt, W

The watt is the transfer of one joule per second.
  The power of an electric device is the product of the potential difference between its ends and the current passed.

Power = potential difference × current
$P = VI$ Unit: watt, W

The specific thermal capacity of a substance is the energy needed to raise one kilogram by one degree.

Specific thermal capacity = energy/(mass × temperature change)
$c = W/m\triangle T$ Unit: joule per kilogram per degree, J kg^{-1} °C^{-1}
                          (or joule per kilogram per kelvin, J kg^{-1} K^{-1})

The specific thermal capacity of water is 4200 J kg^{-1} °C^{-1} (4200 J kg^{-1} K^{-1}).

A 100 W 'heater' in 100 g of water is used for 100 s. What temperature rise results?   [112]

A 250 V 'heater' takes 20 A. It is used in 0.5 t of oil for 20 minutes. Find the temperature rise. The specific thermal capacity of the oil is 2 kJ kg⁻¹ °C⁻¹. [270]

A 10 kW electric crane is 25 per cent efficient. How high can it raise a 500 kg load in 20 s? [422]

A 100 kg block falls 50 m into 100 kg of water. Half the energy raises the water temperature. By how much is the temperature raised? [490]

5.4 Some more questions

1. *A force acts on an object as follows, starting from rest.*
   (i) *It is constant at 10 N for 2 m.*
   (ii) *It rises to 20 N over 2 m.*
   (iii) *It remains at 20 N for 2 m.*
   (iv) *It falls to zero over 2 m.*
   (a) *Plot the force–distance graph.* (b) *From this, find the work done.* [113]
2. (a) *A 150 t aircraft flies at 10 km.* (b) *A 0.1 g hailstone floats at 10 km. What is the potential energy of each?* [430]
3. *Both aircraft and hailstone (question 2) fall to ground without air friction.* (a) *What is the final velocity of each?* (b) *Why are the answers the same?* [304]
4. *A 100 t train accelerates from rest to 20 m s⁻¹ in 20 s.* (a) *What force does the engine exert on the train?* [274] (b) *What work is done by the engine?* [268]
5. *The answers to the last question assumed 100 per cent efficiency. What would they be if the total efficiency was 25 per cent?* [314]
6. *Ninety-five per cent of the electrical energy used by a 100 W lamp appears as 'heat'. If the lamp is submerged in 500 g of water at 20 °C, how long will it be before the water reaches boiling temperature? Assume that 50 per cent of the 'heat' from the lamp remains in the water.* [280]
7. *A 25 g shot is fired from a gun at 1 km s⁻¹. The barrel is 500 mm long. Find* (a) *the shot's acceleration,* (b) *the force from the gun,* (c) *the energy given to the shot.* [105]

5.5 Objectives

Note: The first ten objectives are from Level 1.
When you have studied this chapter, you should be able to

(1) relate work done to area under a force–distance graph;
(2) give examples of conversion of energy from one form to another;

46

(3) define and calculate efficiency in terms of energy input and output;
(4) state that power is the energy transfer in unit time;
(5) state that the unit of power is the watt;
(6) define specific thermal capacity;
(7) solve problems associated with specific thermal capacity;
(8) state examples of electric energy being used for its 'heating' effect;
(9) state that the power produced in an electric circuit is the product of potential difference and current;
(10) calculate the power produced in simple electric circuits;
(11) calculate the energy transferred by a constant force acting over a distance;
(12) solve problems involving constant and variable forces by deducing area under force–distance graphs;
(13) define potential energy and discuss examples;
(14) derive and use $W = mgh$;
(15) use the law of constant energy to discuss situations and to solve problems where mechanical energy is thought to be conserved.

Part 2

Properties of matter: particles and energy

Chapter 6

The particles of matter

6.1 The kinetic theory

Matter includes all solids, liquids and gases. The idea that it is made of tiny particles goes back many hundreds of years. About 300 years ago the belief became widespread. Then scientists started to do tests which helped them accept it.

The modern view is set out in the '**kinetic theory**' of matter. This states the following.

(i)   All matter consists of particles.
(ii) The particles tend to attract each other.
(iii) The particles are always moving. (They have energy.)
(iv) 'Temperature' measures the mean energy of the particles.

The last statement is dealt with again in Chapters 8 and 9. Now we are concerned with the first three.

What *are* these '**particles**'? Here is what we should know now.

An 'atom' is the smallest particle of an element that can exist on its own.

There are about a hundred elements. Each substance is made from one or more of them. (In the same kind of way, there are 26 letters in English. They make up all the thousands of English words.)

A 'molecule' is a group of atoms bonded together, which is able to exist on its own.

Here we shall use the word 'particle' to mean either 'atom' or 'molecule'. The word also covers the units building up solids; often these are neither atoms nor molecules.

How do we know that matter consists of particles? Here is some evidence.

(i) Many solids exist in **crystal** form. The crystals of a given substance always have the same special pattern. This follows from how the particles attract each other and bond together. Solid **metals** consist of jumbled groups of crystals.

(ii) A small volume of a strongly smelling gas released in a corner can soon be detected throughout the room. This spreading or **diffusion** of gases (and liquids) is best explained using fast-moving particles.

(iii) With high magnification, little bits of dirt in water are seen moving around at random. The same is true of little bits of smoke in air. This effect is **Brownian motion**. The little bits are pushed to and fro by the even smaller molecules of the water or the air.

6.2 Forces and their effects on matter

Force tends to change an object's momentum. (See Section 1.3.)

What if the object is fixed, and does not move when pushed?

Fig. 6.1 Trying to move an immovable object

50

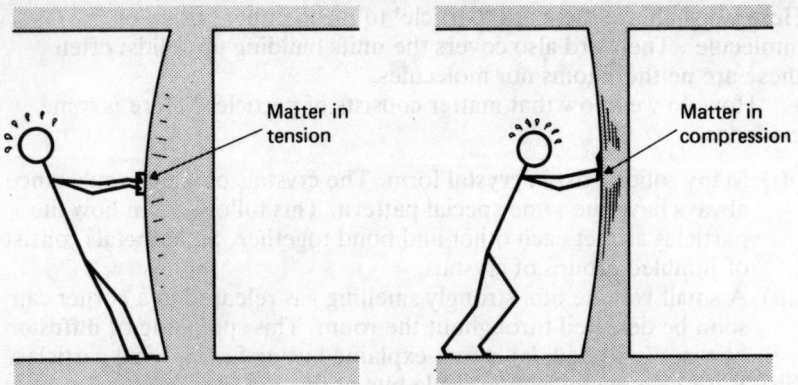

Fig. 6.2 An applied force can produce **tension** or **compression**

The object pushes back with the same force. This follows from Newton's Third Law (section 1.3.).

Where does this '**reaction**' come from? It comes from the particles in the object. Their normal pattern is disturbed by the applied force; the reaction force appears as they try to return to their usual places.

When the normal pattern of the particles is disturbed, the object changes shape. (Often the change is very small indeed.) This is called **distortion**.

A force can distort matter.

Matter can be distorted in two main ways. They are shown in Fig. 6.2.

(There can also be **bending** and **twisting** effects. These are more complex; we shall not deal with them here.)

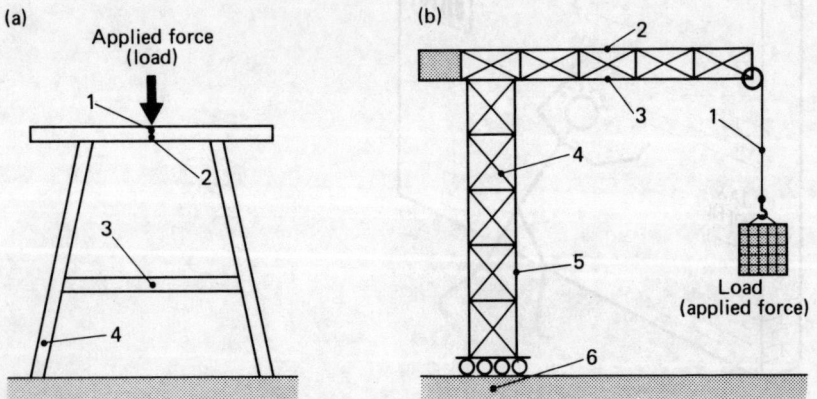

Fig. 6.3

Matter is in tension if the particles are further apart than normal.

Matter is in compression if the particles are closer than normal.

Fig. 6.3 shows two loaded structures. Is the matter in tension or compression at each numbered point?   (a) [114] (b) [293]

6.3 Hooke's law

Three hundred years ago Robert Hooke wanted to know how matter in tension behaves. He did simple tests like the following.

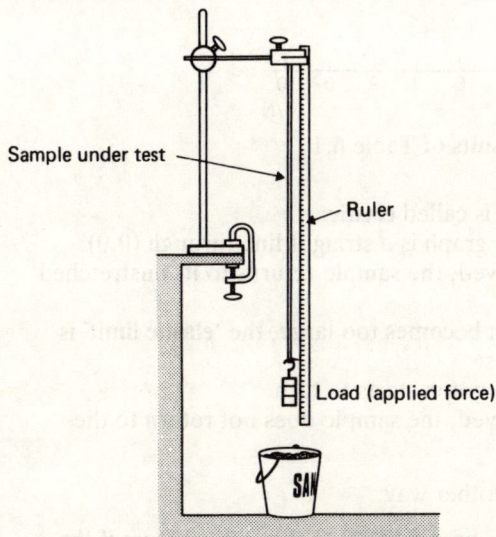

Sample under test

Ruler

Load (applied force)

SA

Fig. 6.4 Testing a sample for Hooke's law behaviour

The weight of the load equals the **tension** T in the sample. As we increase the weight, and thus the tension, the sample becomes longer. The total increase in length, the **extension** $\triangle l$, is measured each time. When the weight is removed, the sample will return to its original length.

The results could be like those in Table 6.1. Plotted on a graph they give a straight line through (0, 0) Fig. 6.5.

We would obtain the same results each time we test the same sample. A straight line extension – tension graph always appears.

Table 6.1

T/N	0	1	2	3	4	5	6	7	8	9	10
$\triangle l$/m	0	0.05	0.10	0.15	0.20	0.25	0.30	0.35	0.40	0.45	0.50

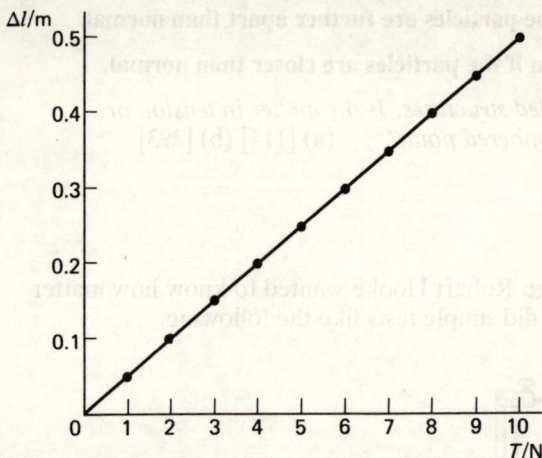

Fig. 6.5 A graph of the results of Table 6.1

This kind of behaviour is called **elastic**:
(i) The extension–tension graph is a straight line through (0,0).
(ii) When the load is removed, the sample returns to its unstretched state.
However, if the tension becomes too large, the '**elastic limit**' is passed. The picture changes:
(i) The extension–tension graph starts to turn.
(ii) When the load is removed, the sample does not return to the unstretched state.

Hooke's law puts all this another way.

The distortion of a sample is proportional to the applied force if the elastic limit is not passed.

All solids obey Hooke's law. (So, indeed, do fluids; these are not so easy to test.) However, in most cases the elastic range is very small. In fact, no common substance shows all the effects well. This is why the Hooke's law test is often done with a helical spring.

The end of a lorry spring moves 80 mm when a 1 t load is put in the back. What will the deflection be after a fifth of the load is removed?   [480]

The cords in Fig. 6.6 are all the same. In (a) the four cords each stretch by 50 mm. By how much will the cord in (b) extend?   [269]

So far, the idea of elastic limit has been used rather loosely. Look at Fig. 6.7.
OP is a straight line. It shows Hooke's law – true elastic behaviour. P is the **proportional limit**. Now the sample is loaded to a

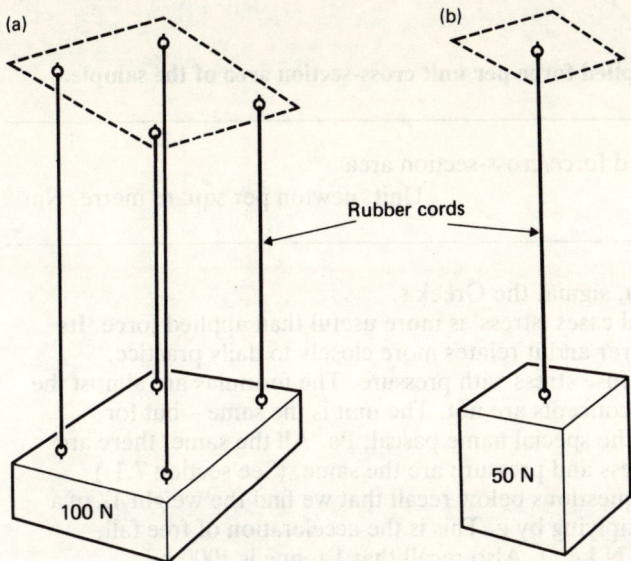

Fig. 6.6

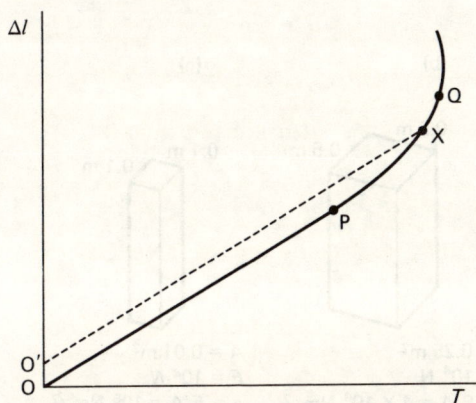

Fig. 6.7 An extension–tension graph for a sample stretched too much

point on PQ, such as X. When this load is removed, the extension will reduce, not to 0 but to OO'. The sample will then be elastic over the range O–X.

If the sample is taken past Q, this no longer happens. Now it is **plastic**; the particles in the sample begin to slip past each other. We come back to this in section 6.7.

Q is the true elastic limit. Often it is very close to P, the proportional limit, but it is *not* the same.

6.4 Stress

Stress is the applied force per unit cross-section area of the sample.

Stress = applied force/cross-section area
$\sigma = F/A$ Unit: newton per square metre, Nm^{-2}

The symbol is σ, sigma, the Greek s.

In most real cases 'stress' is more useful than applied force. Its meaning is clearer and it relates more closely to daily practice.

Do not confuse stress with pressure. The formulas are almost the same – but the concepts are not. The unit is the same – but for pressure it has the special name pascal, Pa. All the same, there are cases where stress and pressure are the same. (See section 7.1.)

To do the questions below recall that we find the weight W of a mass m by multiplying by g. This is the acceleration of free fall, 10 m s^{-2} (or 10 N kg^{-1}). Also recall that 1 tonne is 1000 kg.

1 t loads are applied to the blocks shown in Fig. 6.8. What is the compressive stress in each?

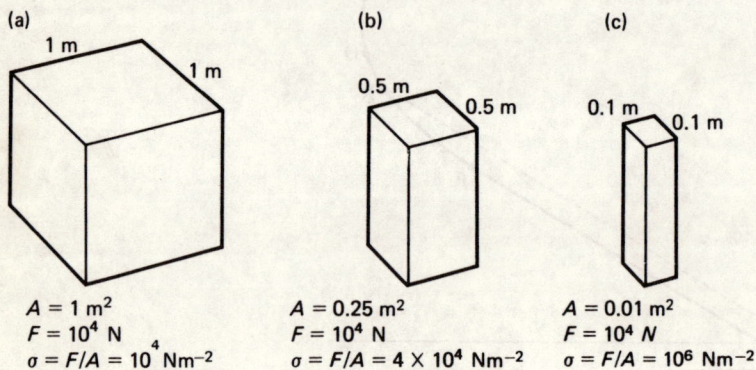

(a) 1 m, 1 m
(b) 0.5 m, 0.5 m
(c) 0.1 m, 0.1 m

$A = 1$ m^2 $A = 0.25$ m^2 $A = 0.01$ m^2
$F = 10^4$ N $F = 10^4$ N $F = 10^4$ N
$\sigma = F/A = 10^4$ Nm^{-2} $\sigma = F/A = 4 \times 10^4$ Nm^{-2} $\sigma = F/A = 10^6$ Nm^{-2}

Fig. 6.8

We see that the applied forces are the same – but stress depends on cross-section area.

A building has a mass of 1000 t. It is carried by ten solid steel piles of radius 100 mm. What is the compressive stress in each?

For *one* pile: $F = 1000 \times 1000 \times 10/10$ N $= 10^6$ N
$A = \pi \times 0.1 \times 0.1$ m$^2 = \pi \times 10^{-2}$ m^2 $\sigma = ?$
$\sigma = F/A = 10^6/(\pi \times 10^{-2})$ Nm$^{-2} = 3.2 \times 10^7$ Nm^{-2}

A 10 t elephant stands on one leg. The smallest section of thigh bone is 10^4 mm². What is the highest compressive stress in the bone? [321]

A crane lifts 20 t with a cable of 80 mm diameter. What is the tension in the cable? What is the tensile stress in the cable? [412]

6.5 Strain

Stress is more useful than applied force. In the same way **strain** is more useful than 'distortion'. Its meaning is clearer and it relates better to daily practice. Again the reason is that strain, unlike distortion, can be discussed for any *substance* rather than for given *samples*.

Strain is change of dimension per unit original dimension.

For instance   Strain = change of length/normal length

$\epsilon = \triangle l/l$

The symbol is ϵ, epsilon, the short Greek e. Strain has no unit – it is a ratio of lengths in this case.

Each cord in Fig. 6.9 is stretched by 100 mm. What is the tensile strain in each?

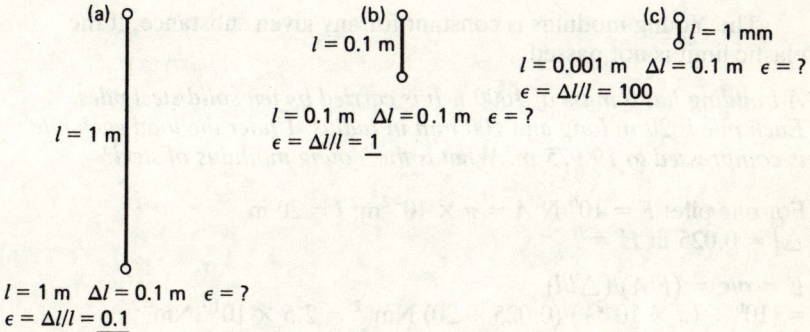

(a) $l = 1$ m

(b) $l = 0.1$ m

$l = 0.1$ m $\Delta l = 0.1$ m $\epsilon = ?$

$\epsilon = \Delta l/l = \underline{1}$

(c) $l = 1$ mm

$l = 0.001$ m $\Delta l = 0.1$ m $\epsilon = ?$

$\epsilon = \Delta l/l = \underline{100}$

$l = 1$ m $\Delta l = 0.1$ m $\epsilon = ?$

$\epsilon = \Delta l/l = \underline{0.1}$

Fig. 6.9

One steel pile of a building was originally 20 m long. Under load it is compressed to 19.75 m. What is the compressive strain?

$l = 20$ m $\triangle l = 0.25$ m $\epsilon = ?$

$\epsilon = \triangle l/l = 0.25/20 = 0.0125$

The femur (thigh bone) of an elephant is 1.5 m long. When the animal stands on one leg the femur is 10 mm shorter. What is the compressive strain? [330]

A crane cable is 50 m long. When lifting a load, the tensile strain in it is 0.002. By how much does it stretch? [278]

6.6 Elastic modulus

'Elastic modulus' (modulus = constant) often appears in lists of the main properties of matter. There are a number of elastic moduli; each concerns one kind of situation. However, each is defined as the ratio of a stress to a strain. And they all describe how elastic the substance is.

The Young modulus is the one we use here. It is the ratio of the stress and the strain we have used in this chapter. It is named after Thomas Young; he worked on this subject at the start of the nineteenth century.

The Young modulus of a substance is the tensile or compressive stress per unit linear strain.

Young modulus = tensile or compressive stress/linear strain

$$E = \frac{F/A}{\triangle l/l} \qquad \text{Unit: newton per square metre, N m}^{-2}$$

The Young modulus is constant for any given substance, if the elastic limit is not passed.

A building has a mass of 1000 t. It is carried by ten solid steel piles. Each pile is 20 m long and 100 mm in radius. Under the load each pile is compressed to 19.975 m. What is the Young modulus of steel?

For *one* pile: $F = 10^6$ N $A = \pi \times 10^{-2}\text{m}^2$ $l = 20$ m
$\triangle l = 0.025$ m $E = ?$

$E = \sigma/\epsilon = (F/A)/(\triangle l/l)$
$= (10^6 \div (\pi \times 10^{-2}))/(0.025 \div 20)$ Nm^{-2} $= 2.5 \times 10^{10}$ Nm^{-2}

The smallest section of a 10 t elephant's 1.5 m femur is 10^4 mm^2. When the animal stands on one leg, the bone shortens by 10 mm. What is the Young modulus of the bone? [346]

A crane cable is 50 m long and 80 mm in diameter. When lifting 20 t it stretches by 0.1 m. What is the cable's Young modulus? [437]

What does the Young modulus mean in practice? We could test

special samples of a number of materials. Each sample would have the same length and cross-section area.

Then $E \propto F/\triangle l$

If E is low, a small force gives a large change in length. Such samples have high **elasticity**. We can also say they have low **stiffness**.

If E is high, a large force gives a small change in length. Such samples have low elasticity, or high stiffness.

Check these points in Table 6.2.

Table 6.2 The elasticity of different materials

Material	Young modulus E/Nm^{-2}	Elasticity	Stiffness
Tungsten	4×10^{11}	Very low	Very high
Steel	2×10^{11}	Low	High
Perspex	6×10^9	High	Low
Polyisoprene (rubber)	5×10^6	Very high	Very low

6.7 Fracture

Scientists and engineers need to know more about a substance than how elastic it is. A sample stressed more and more behaves in different ways. Refer to fig. 6.10.

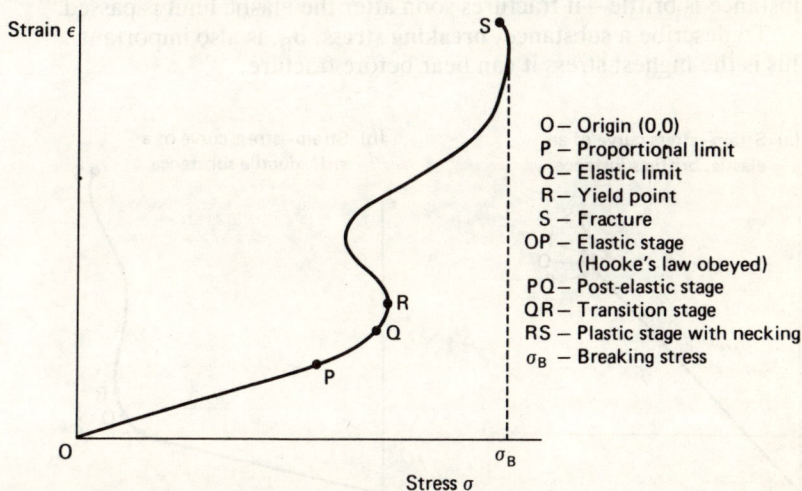

O – Origin (0,0)
P – Proportional limit
Q – Elastic limit
R – Yield point
S – Fracture
OP – Elastic stage (Hooke's law obeyed)
PQ – Post-elastic stage
QR – Transition stage
RS – Plastic stage with necking
σ_B – Breaking stress

Fig. 6.10 A typical strain–stress graph

(i) **The elastic state** (Hooke's law region) (OP)
The sample deforms in proportion to the stress. When the stress is removed, the sample returns to the unstretched state. This Hooke's law behaviour ends at the proportional limit.

(ii) **The post-elastic stage** (PQ)
The sample deforms in proportion to stress, but less than before. When the stress is removed, the sample remains slightly strained. (We say it has a permanent 'set'.) The limit of this behaviour is the elastic limit.

(iii) **The transition and plastic stages** (QR and RS)
The sample structure starts to change. Layers of particles slip past each other. At one or more points the section area becomes less. This effect is called '**necking**'; the sample deforms plastically. If the stress is removed, only the elastic deformation goes. QR in the graph represents the transition from elastic to plastic behaviours.

Plastic behaviour starts at the **yield point**. It goes on until the sample breaks, or **fractures**.

Except in detail, all solid samples will have a strain – stress curve like this.

We have seen that samples differ in their **elasticity**, or **stiffness**. If the slope of OP is high, the sample is elastic (low stiffness). If the slope is low, the sample is stiff (low elasticity).

Solid samples also differ in their **brittleness** or **ductility**. If RS in the graph is long, the substance is ductile. This means it can be stretched to a great length as a wire or thread. If RS is short, the substance is brittle – it fractures soon after the elastic limit is passed.

To describe a substance, **breaking stress**, σ_B, is also important. This is the highest stress it can bear before fracture.

(a) Strain–stress curve of an elastic, brittle substance

(b) Strain–stress curve of a stiff, ductile substance

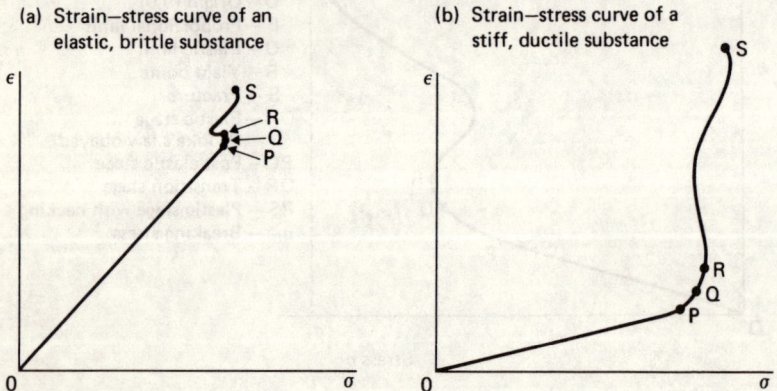

Fig. 6.11

We can test a metal (in the form of a wire) to fracture and explore the shape of the strain-stress curve. We can thus determine the Young modulus, the elastic limit and the breaking stress of the wire.

This can be done using a piece of equipment similar in principle to that in Fig. 6.4. Increasing loads are added to the wire and the increase in length measured. Even large loads give only small increases in length which require to be measured with a vernier gauge. Before stretching, the diameter of the wire must be measured at several places along its length, using a micrometer, and an average value determined. The length of wire is measured with a ruler.

The stress (F/A) and strain ($\triangle l/l$) can be determined given that F is the load in Newtons, A is the cross-section of the wire in m^2, l is the original length of the wire and $\triangle l$ the change in length under the applied load (both in m). From the linear portion of the resulting stress-strain curve we can obtain the Young modulus (E) which is equal to the slope.

$$E = \frac{F/A}{\triangle l/l}$$

DANGER – Always put a safety screen in front of your experiment to ensure that when the wire breaks it cannot snap out at you.

It is useful to repeat the experiment but only along the linear part of the curve. Then remove the load continuously and determine the corresponding *decrease* in length. Hence determine whether or not the wire is elastic, i.e. whether the decrease in length with load is the reverse of the increase.

6.8   Some more questions

1. *Sketch the strain–stress curve of an elastic, ductile substance. Show on it these points: elastic limit, fracture point, proportional limit, yield point. Explain those terms.*   [445]
2. *A straight 100 mm human hair was of radius 0.10 mm. It broke when loaded to 100 g, having stretched by 1.0 mm. What is the ratio of breaking stress to breaking strain?*   [305]
3. *A cube of wood of dimensions 100 mm is cut from a balsa tree. It is loaded in the three directions. The details are shown. Complete the table.*

Table 6.3 Stresses and strains for a wood sample

Direction	Applied force/N	Compression/mm	Stress/Nm^{-2}	Strain E/MNm^{-2}
Longitudinal	10^6	1.67		
Radial	10^5	3.33		
Transverse	10^4	1.00		

Relate the directions to a tree trunk; explain why the Young modulus is not the same in each case. [296]

4. *Complete the table.*

Table 6.4 Stretching wires [283]

Metal	E	Sample length	Sample diameter	Load	Sample extension
Aluminium	1 unit	3 units	2 units	50 units	
Chromium	4 units	1 unit	1 unit	50 units	1 unit

6.9 Objectives

Note: The first four objectives are from Level 1.
When you have studied this chapter, you should be able to

(1) outline the 'kinetic theory of matter';
(2) define atoms and molecules;
(3) discuss evidence for the particle model of matter;
(4) state that forces can deform objects;
(5) recognise tensile and compressive forces;
(6) state Hooke's law, describe how to demonstrate it, and solve simple problems using it;
(7) define stress and calculate it for given cases;
(8) define strain and calculate it for given cases;
(9) draw extension–force and strain–stress graphs for a elastic material extended up to and beyond its elastic limit;
(10) define the Young modulus and relate it to the stiffness of a material;
(11) solve problems involving stress, strain and the Young modulus;
(12) describe and discuss the form of strain–stress graphs for brittle and ductile materials;
(13) describe elastic and plastic deformation and elastic limit;
(14) describe how to test a metal wire to destruction in order to plot its strain-stress curve.

Chapter 7

The elastic behaviour of a gas

7.1 Stress and pressure

In the last chapter we looked at the effects of forces on **solids**. The two main ones were tension (stretching) and compression (squashing). Here the particles are pulled further apart than normal, or pushed closer together than normal.

The results of both were described in terms of **strain**. This is the change of dimension of unit dimension of the solid sample. (For instance, it can be the change of length of unit length.)

The applied force was described in terms of **stress**. Stress is the force acting on unit cross-section area of the solid sample.

Solids differ from other kinds of matter – liquids and gases – in one major respect. Solids cannot flow; liquids and gases can. We use the term **fluid** to describe the liquid and gas states of matter; this means something that can flow.

Figure 7.1 shows how the three states of matter differ. In the sketches we show their particles (in two dimensions).

In a solid the particles are fixed in place; in a fluid the particles are free to move.

We shall see the reasons for all this later, in Chapter 9. Now we just accept that because fluids can flow, forces act on them differently.

To find out how forces affect solids we could use a device like the Hooke's law test rig in Fig. 6.4. We should be able to measure stress

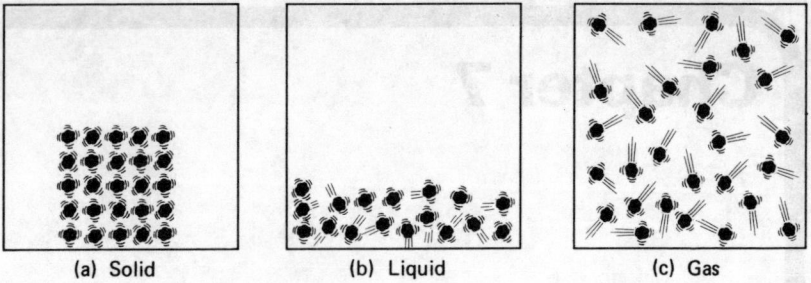

Fig. 7.1 The particles of matter

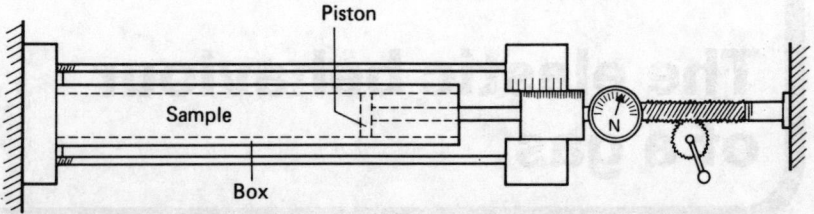

Fig. 7.2 Rig to measure tensile stress and strain for a fluid sample

and strain more accurately, and test plastic behaviour and fracture, too, if we wish.

Because fluids flow, we cannot do the same with them. The sample would need to be trapped in a box with a piston. We could try something like that in Fig. 7.2. But testing the sample would *not* be easy!

One problem would be leakage – the sample could flow past the piston; so could air. A more serious problem is the force applied by the sample on the box. (And, because of Newton's third law, the force from the box on the sample.) In the case of gases the biggest problem of all is that their particles have no **cohesion** – in effect, they do not attract each other, so there is no such thing as tensile stress for a gas.

Instead of stress we use **pressure** with fluids. We saw in section 6.5 that the two are similar but not the same. However, the concept of pressure *can* also be used where solid surfaces act on each other; the next two worked examples show this.

Pressure is the applied force per unit surface area.

Pressure = applied force/surface area

$p = F/A$ Unit: pascal, Pa ($\equiv Nm^{-2}$)

The SI **unit of pressure** is the newton per square metre, just as with stress. For pressure, however, it has the special name 'pascal' (Pa).

A 5 t crate is 1 m × 5 m × 750 mm. It stands on the three different faces in turn. What pressure is applied on the ground in each case?

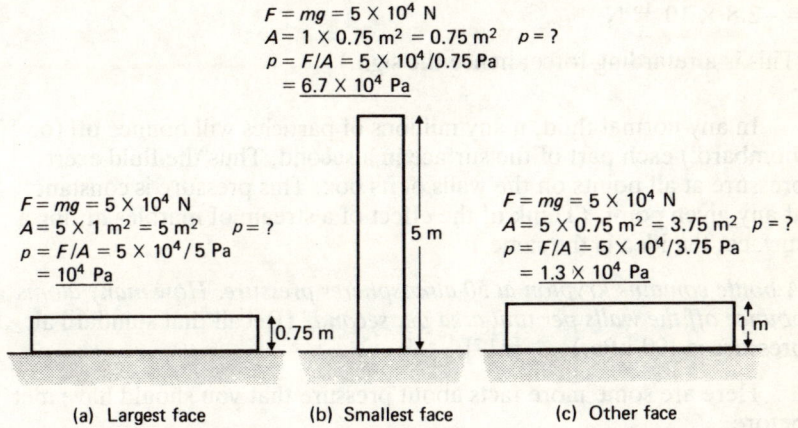

$F = mg = 5 \times 10^4$ N
$A = 1 \times 0.75$ m² $= 0.75$ m² $p = ?$
$p = F/A = 5 \times 10^4/0.75$ Pa
  $= 6.7 \times 10^4$ Pa

$F = mg = 5 \times 10^4$ N
$A = 5 \times 1$ m² $= 5$ m² $p = ?$
$p = F/A = 5 \times 10^4/5$ Pa
  $= 10^4$ Pa

$F = mg = 5 \times 10^4$ N
$A = 5 \times 0.75$ m² $= 3.75$ m² $p = ?$
$p = F/A = 5 \times 10^4/3.75$ Pa
  $= 1.3 \times 10^4$ Pa

5 m

0.75 m

1 m

(a) Largest face   (b) Smallest face   (c) Other face

Fig. 7.3

A 10 t elephant stands on one foot. The area touching the ground is 2.5×10^{-2} m². What is the pressure on the ground?   [243]

Air exerts a pressure of about 100 kPa. This figure is 'standard air pressure' (sometimes called an 'atmosphere'). The actual value varies from place to place and from minute to minute; it depends mainly on the weather and the height above sea level. **We measure air pressure with barometers.**

An adult's skin has an area of 2 m². (a) What force does the air exert on it? (b) How many elephants does this correspond to?   [340]

What is the cause of the pressure in a fluid?
Consider the particles of a fluid, a liquid or a gas. (Look at Figs. 7.1(b) and (c).)

(i) Fluid particles are moving around all the time.
(ii) So, fluid particles have momentum. (See section 1.3.)
(iii) So, as they bounce off the walls of the box they apply a force on the walls.

We can combine Newton's second and third laws to help us think about this. (Check by refering to section 2.1 if necessary.)

The force applied equals the change of momentum in unit time.

A krypton atom has a mass of 1.4×10^{-25} kg. It hits a surface at 90° at 1 km s⁻¹ and is bounced back without loss of energy in 1 μs. What force does it exert?

$m = 1.4 \times 10^{-25}$ kg $v_1 = 10^3$ m s⁻¹ $v_2 = -10^3$ m s⁻¹
$t = 10^{-6}$ s $F = ?$
$F = m(v_2-v_1)/t = 1.4 \times 10^{-25} (-10^3-10^3)/10^{-6}$ N
$= -2.8 \times 10^{-16}$ N

(This is a retarding force; hence the sign.)

In any normal fluid, many millions of particles will bounce off (or 'bombard') each part of the surface in a second. Thus the fluid exerts pressure at all points on the walls of its box. This pressure is constant at any given point. (Think of the effect of a stream of marbles hitting a surface; the idea is the same.)

A bottle contains krypton at 50 atmospheres pressure. How many atoms bounce off the walls per unit area per second? (Recall that standard air pressure is 100 kPa.)   [417]

Here are some more facts about pressure that you should have met before.

The pressure at a point in a fluid is the same in all directions. If this were not the case, the particles would shift until it became so.

The pressure in a fluid does not depend on the shape of the box. This follows from the previous statement.

Pressure acts at 90° to any surface in contact. By thinking about the particles' momenta, we can soon accept this for fluid pressure. It is also true for the pressure on a surface from a solid in contact – but that is not so easy to prove.

The pressure at a depth in a fluid equals the weight of fluid above unit area. We can compare this with the first problem in this chapter. You may know that we use this idea and a sketch like Fig. 7.4 to show that **the pressure at a depth in a fluid is the product of the depth, the mean density and the acceleration of free fall.**

We recall that **the density of a substance is the mass of unit volume.** The unit is the kilogram per cubic metre.

Pressure in a fluid = depth × density × g

$p = h\rho g$ Unit: pascal, Pa

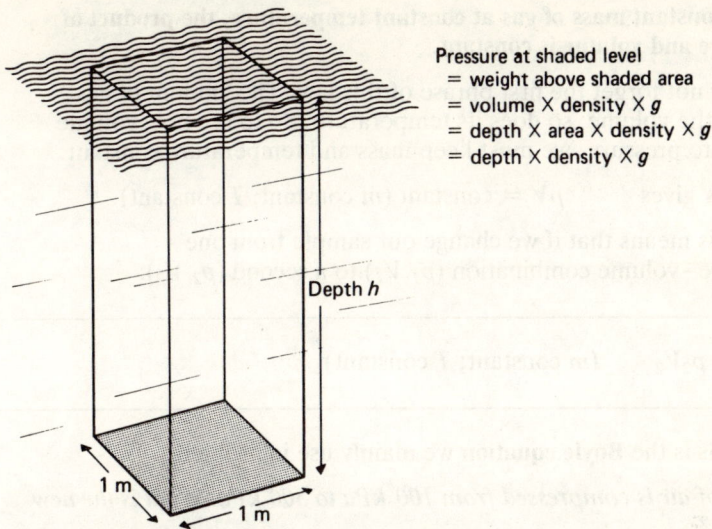

Pressure at shaded level
= weight above shaded area
= volume × density × g
= depth × area × density × g
= depth × density × g

Depth h

1 m

1 m

Fig. 7.4 $p = h\rho g$

A cylindrical oil tank has radius 5 m. It is filled 3 m deep with oil of density 800 kg m^{-3}. Use two different methods to show that the oil applies a force of about 1.9 MN on the tank base.

(a) Using $F = pA = h\rho gA$:
   $h = 3$ m $\rho = 800$ kg m^{-3} $g = 10$ m s^{-2}
   $A = \pi \times 5^2$ m^2 $F = ?$
   $F = h\rho gA = 3 \times 800 \times 10\,\pi \times 5^2$ N $= 1.9$ MN.
(b) Using $F = W = V\rho g$:
   $h = 3$ m $A = \pi \times 5^2$ m^2 $\rho = 800$ kg m^{-3}
   $g = 10$ m s^{-2} $F = ?$
   $F = V\rho g = 3 \times \pi \times 5^2 \times 800 \times 10$ N $= 1.9$ MN

Standard air pressure is 100 kPa. The density at sea level is 1.25 kg m^{-3}. If the air density were constant, how deep would the atmosphere be? [290]

7.2 Boyle's law

How liquids behave when stressed is not important to us at the moment. However, we need to deal very fully with **gases**.

How does a gas deform with change of pressure? The answer is given by **Boyle's law**; this is over 300 years old.

For a constant mass of gas at constant temperature, the product of pressure and volume is constant.

Do not forget the first phrase of this law. The amount of gas affects the volume; so does its temperature. So to see how volume relates to pressure, we must keep mass and temperature constant.

The law gives pV = constant (m constant; T constant).

This means that if we change our sample from one pressure–volume combination (p_1, V_1) to a second (p_2, V_2):

$$p_1V_1 = p_2V_2 \qquad (m \text{ constant}; T \text{ constant})$$

This is the **Boyle equation** we mainly use in problems.

10 m³ of air is compressed from 100 kPa to 500 kPa. What is the new volume?

$V_1 = 10 \text{ m}^3 \; p_1 = 10^5 \text{ Pa} \; p_2 = 5 \times 10^5 \text{ Pa} \; V_2 = ?$

$p_1V_1 = p_2V_2 \rightarrow V_2 = p_1V_1/p_2 = 10^5 \times 10/5 \times 10^5 \text{ m}^3 = 2 \text{ m}^3$

10 l of argon at five atmospheres pressure is compressed to 1 l. Find the new pressure.   [115]

Notes:
(1) Like the tonne (1000 kg), the **litre** (l, 10^{-3} m³) and the **atmosphere** (100 kPa) are not SI units. They are widely used, however, so we must know them.
(2) In problems about gases, if mass and temperature are not mentioned, we assume they are constant.

A gas sample is doubled in volume. What happens to the pressure?   [423]

7.3 Charles' law

What happens if the gas temperature is *not* constant?

We know that **most substances expand with temperature rise**. (One common one does not – **water** between 0 °C and 4 °C; when warmed through that range the liquid contracts.)

All gases expand with temperature rise. In other words, the volume of warmed gas samples becomes greater. This is a kind of deformation, so we must think about it here.

The effect is most clearly covered by **Charles' law**. (This is named after the French professor Jacques Charles, who first noted it almost 200 years ago. He made the first 'flight' in a hydrogen balloon.)

Charles' law relates the volume and temperature of a gas sample. This time we keep the pressure constant.

For a constant mass of gas at constant pressure, the volume is proportional to the absolute temperature.

The concept of **absolute** (or **'thermodynamic'**) **temperature** may be new. What is it?

If we measure the volume of a gas sample at different celsius temperatures, we get a graph like that in Fig. 7.5.

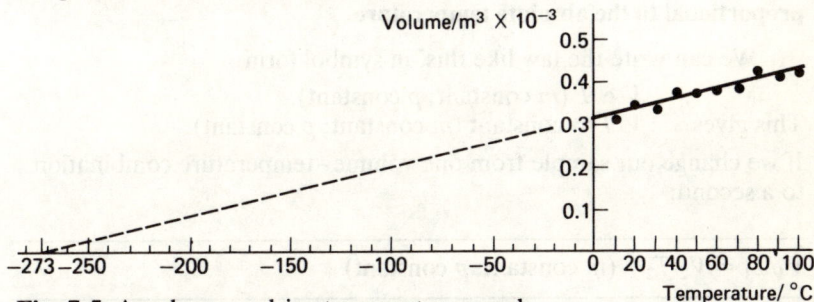

Fig. 7.5 A volume–celsius temperature graph

Clearly this is a straight-line graph – but it does *not* pass through (0,0). However, if we produce the line back, below 0 °C, it will reach the temperature axis at some point.

Whatever gas we test, the volume–temperature graph cuts the temperature axis at about −273 °C.

This lowest temperature is called the **absolute zero** of temperature. From the graph we could say that it is the temperature at which any gas would have no volume. (Strictly this is not true in fact.)

If this point is made the origin of the volume–temperature graph, we have a straight line passing through (0, 0). We must 'invent' a new temperature scale, whose zero is at −273 °C. The **absolute temperature** scale is used. It is often called the **thermodynamic scale.**

The symbol for absolute temperature is *T*. Its unit has the same size as the degree celsius. It is sometimes called the *degree absolute*; the SI name is the **kelvin** (K).

Thus   0 K = −273 °C;
          273 K = 0 °C   (the ice temperature);
          373 K = 100 °C   (the steam temperature).

The ice and steam temperatures are the most common '**fixed temperatures**' used for making and testing thermometers. (See Chapter 9.)

Here are some more fixed temperatures. Convert those on the celsius scale to absolute values, and vice versa.

(i)   The freezing temperature of gold, 1064 °C.
(ii) The freezing temperature of silver, 1235 K.
(i)   The freezing temperature of zinc, 419.6 °C.
(ii) The triple point of oxygen, 54.4 K.
(iii) The boiling temperature of hydrogen, −253 °C.
(iv) The triple point of hydrogen, 14 K.   [281]

Now we can return to **Charles' law!**

For a constant mass of gas at constant pressure, the volume is proportional to the absolute temperature.

We can write the law like this, in symbol form.

$$V \propto T \quad (m \text{ constant}; p \text{ constant}).$$

This gives V/T = constant (m constant; p constant).

If we change our sample from one volume–temperature combination to a second:

$V_1/T_1 = V_2/T_2$ (m constant; p constant)

This is the **Charles' equation** we find most useful. When we use it, the temperature must be absolute, given in kelvin (K); also mass and pressure must be constant.

The absolute temperature of a neon sample doubles. The volume therefore . . .   [228]

A balloon is filled with 100 m³ of hydrogen at 27 °C. At a certain height, the air temperature is −23 °C. How large is the balloon there?

$V_1 = 100 \text{ m}^3 \ T_1 = 300 \text{ K} \ T_2 = 250 \text{ K} \ V_2 = ?$
$V_1/T_1 = V_2/T_2 \rightarrow$
$V_2 = V_1 T_2/T_1 = 100 \times 250/300 \text{ m}^3 = 83.3 \text{ m}^3$

A gas sample at −123 °C is warmed until its volume has tripled. To what temperature is it warmed?   [116]

7.4 The pressure law

We have used two gas laws so far. Here they are in short form.

Boyle's law $V \propto 1/p$ (m constant, T constant).
Charles' law V/T = constant   (m constant, p constant).

A third gas law must exist, to relate the pressure and temperature of samples. It would have the volume constant.

This third law is the **pressure law**.

For a constant mass of gas at constant volume, the pressure is proportional to the absolute temperature.

Pressure law $p \propto T$ (m constant, V constant).

When we change our gas sample from one pressure–temperature combination to a second:

$p_1/T_1 = p_2/T_2$ (m constant, V constant)

Notes: Again we use absolute temperatures, in kelvin (K).

A cylinder contains krypton at 50 atmospheres. In a fire, the temperature rises from 27 °C to 627 °C. Find the new gas pressure. [161]

The air in a car tyre at 27 °C is pumped from standard pressure to 250 kPa. What temperature could be reached? [119]

7.5 The ideal gas equation

The three gas laws we have met are strictly true only for ideal, or perfect, gases. In fact we say that **an ideal gas is one which obeys the gas laws!** *No* real gas is truly ideal; however, all are near enough as long as the pressure is not too high and the temperature is not too low. At normal temperatures and pressures this applies to all of the gases which are elements, and thus to dry air with no carbon dioxide content. The nature of an ideal gas is explained on a particle basis in section 7.6.

If we combine the three gas law equations, we obtain the **ideal gas equation**.

$p_1V_2/T_1 = p_2V_2/T_2.$ (m constant)

Clearly the three gas equations are inside this combined one. The ideal gas equation relates the pressure, volume and absolute temperature of a sample of constant mass. If we keep any one of those three related measures constant, we get back to one of the three gas laws.

Because of this last fact, we need never use the gas law equations in problems. The next three questions explain this.

Some air is blown into a toy balloon, giving it a volume of 0.75 l at 27 °C. The balloon is left near a fire. It expands to 1.5 l before bursting. What is its final temperature?

$V_1 = 0.75 \times 10^{-3}$ m^3 $T_1 = 300$ K $V_2 = 1.5 \times 10^{-3}$ m^3
$T_2 = ?$ $p_1 = p_2$
$p_1 V_1/T_1 = p_2 V_2/T_2 \rightarrow$
$T_2 = V_2 T_1/V_1 = 1.5 \times 10^{-3} \times 300/0.75 \times 10^{-3}$ K = 600K

200 m^3 of air is pumped into a 0.5 m^3 container without change of temperature. What is the air pressure in the container? [392]

The air container in the last question is now cooled from 27 °C to −173 °C. What is the new pressure? [312]

Here are some questions in which all three related measures vary.

In a chemical reaction 300 cm^3 of gas at 27 °C and 98 kPa is produced. What would be the volume at STP? (STP is 273 K and 100 kPa.)

$V_1 = 0.3 \times 10^{-3}$ m^3 $T_1 = 300$ K $p_1 = 98 \times 10^3$ Pa
$V_2 = ?$ $T_2 = 273$ K $p_2 = 10^5$ Pa
$p_1 V_1/T_1 = p_2 V_2/T_2$
$\rightarrow V_2 = p_1 V_1 T_2/T_1 p_2 = 98 \times 10^3 \times 0.3 \times 10^{-3} \times 273/300 \times 10^5$ m^3
$= 2.68 \times 10^{-4}$ m^3 (268 cm^3)
Note: 1 m^3 = 10^6 cm^3

A tank contains 9 m^3 of a gas at 0.5 MPa and −73 °C. In an explosion and fire the tank is squashed to two − thirds its volume and raised in temperature by 500 °C. What is the gas pressure? [121]

7.6 Other forms of the ideal gas equation

So far we have used the simplest version of the ideal gas equation.

$p_1 V_1/T_1 = p_2 V_2/T_2$ (*m* constant).

For a constant mass of gas, this means that the product of pressure and volume divided by the absolute temperature is constant. Let us call the constant '*K*'.

Then $pV/T = K$.

This gives $pV = KT$.

What is the constant *K*? Tests show that its value depends on the gas concerned and the mass *m* of the sample. To describe each gas here, we use the **specific gas constant**, *r*. Each gas has its own value of *r*.
    The ideal gas equation becomes

$pV = mrT$

It is not easy to define the specific gas constant. First let us find its unit from the equation:

$$r = pV/mT$$

So unit of $r = (Nm^{-2})(m^3)/(kg)(K) = Nm\ kg^{-1}\ K^{-1} = J\ kg^{-1}\ K^{-1}$.

You may know that the $J\ kg^{-1}\ K^{-1}$ (joule per kilogram per kelvin) is also the unit of specific thermal capacity. (See Chapter 9.) Indeed the specific gas constant is related to specific thermal capacity.

We can also find the value of r for any gas. An example follows. We recall that **the density, ρ, of a substance is the mass of unit volume.**

At STP the density of hydrogen is 0.09 kg m^{-3}. Find its specific gas constant.

$$T = 273\ K \quad p = 10^5\ Pa \quad \rho = 0.09\ kg\ m^{-3} \quad r = ?$$

$$pV = mrT \rightarrow r = pV/mT = p/\rho T = 4.1\ kJ\ kg^{-1}\ K^{-1}$$

Sometimes it is important to be able to relate the ideal gas equation to the number of moles (n) present rather than the mass (m). Since $n = m/M$ (where M = relative molecular mass of the gas), $pV = nMrT = nRT$ (R, molar gas constant $= Mr$).

The next form of the ideal gas equation comes from thinking about the real meanings of temperature and pressure. Both involve the motion of the gas particles.

Figure 7.6(a) shows a cube-shaped box of unit volume. It contains n gas particles; the mass of each is m_p. The particles are moving at high speed in all directions. They often collide with each other and with the walls. This causes what we call pressure. Because of the collisions the particles vary in speed.

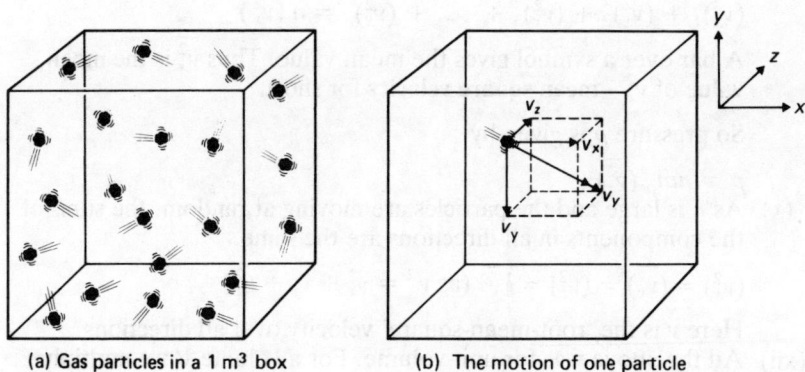

(a) Gas particles in a 1 m³ box   (b) The motion of one particle

Fig. 7.6

Now we derive the new equation.

(i) One particle has velocity v as shown in Fig. 7.6(b). The components of v in the x, y and z directions are v_x, v_y and v_z.

(ii) The particle has momentum $m_p v$, with components $m_p v_x$, $m_p v_y$, and $m_p v_z$.

(iii) The particle hits the right-hand wall; its momentum $m_p v_x$ is reversed; the other momentum components do not change since they are at 90° to *that* wall. This assumes an elastic collision. (See section 4.5.)

(iv) The particle's change of momentum in the x-direction is therefore $2m_p v_x$.

(v) After $2/v_x$ seconds, the particle will hit the right-hand wall again. This assumes that the particle is very small; that it does not collide with other particles; that particles don't affect each other; that collision times are tiny; that collisions are elastic. So the particle will hit the wall $v_x/2$ times per second.

(vi) From Newton's second law, the particle will exert a force F on the wall. The value of F is given by the change of momentum in unit time (per second).
$$F = 2\,m_p v_x \times v_x/2 = m_p v_x^2$$

(vii) The particle will exert a pressure p on the wall. Its value comes from dividing the force F by the area of the wall -1 m^2.
$$p = m_p v_x^2$$

(viii) The *total* pressure on this wall is the sum of the pressure from all n particles.
$$p = m_p \left[(v_x^2)_1 + (v_x^2)_2 + (v_x^2)_3 + \ldots + (v_x^2)_n \right]$$

(ix) We now replace the sum of the squares by the product of n and the mean of the squares. This is a much simpler way to write the sum.
$$(v_x^2)_1 + (v_x^2)_2 + (v_x^2)_3 + \ldots + (v_x^2)_n = n\,(\overline{v_x^2})$$

A bar over a symbol gives the mean value. Thus $\overline{v_x^2}$ is the mean value of v_x^2 – **mean square velocity** for short.

So pressure p is given by
$$p = nm_p\,(\overline{v_x^2})$$

(x) As n is large and the particles are moving at random, the sums of the components in all directions are the same.
$$(\overline{v_x^2}) = (\overline{v_y^2}) = (\overline{v_z^2}) = \tfrac{1}{3}\overline{v^2} \;(\text{as } v^2 = v_x^2 + v_y^2 + v_z^2)$$

Here v is the 'root-mean-square' velocity over all directions.

(xii) All the above was for unit volume. For a volume V we multiply both sides of the above by V.
$$pV = \tfrac{1}{3}\,nVm_p\overline{v^2}$$

(xiii) Lastly we use N, the number of particles in volume V. $N = nV$.

So $pV = \frac{1}{3} Nm_\mathrm{p}\overline{v^2}$

The product of gas pressure and volume is a third of the product of the number of particles, the mass of each particle, and the mean square velocity of the particles.

To obtain the particle form of the ideal gas equation, we assumed certain things. They define an **ideal gas** on a particle basis.

(i) A gas sample consists of very many particles.
(ii) The particles are in constant random motion.
(iii) Forces on and between particles are ignored except when they collide.
(iv) The particles are infinitesimal (infinitely small).
(v) Collisions are elastic.
(vi) Collisions take infinitesimal time.

Not all of these statements are true of real gases. But they are not far wrong – the equation we derived gives results close to practice.

We have also assumed (but not so far stated) that our 'special' particle did not hit others as it bounced between the walls. This won't really be true either; however, as collisions are elastic the statement is fair. This is because during a collision, the two particles simply exchange motions.
(If there is an elastic impact between an object moving at 4 km s^{-1} north and one of the same mass coming south at 2 km s^{-1}, the first will then go south at 2 km s^{-1} while the other moves north at 4 km s^{-1}. Only thus can momentum and energy each be conserved.)

Now we have two new forms of the ideal gas equation.

$$pV = mrT,$$
and $\quad pV = \frac{1}{3} Nm_\mathrm{p}\overline{v^2}$

They show how temperature relates to particle speed.

The absolute temperature of a sample is proportional to the particles' mean square velocity.

We would expect this. If we add energy to a gas the temperature rises. The energy we give to an ideal gas must increase the particles' mean kinetic energy. We saw in section 4.4 that kinetic energy is given by

$$W = \frac{1}{2} mv^2$$

This concept relates to the statement (section 6.1) that
'temperature' measures the mean energy of the particles.
Here is a question using the new equation. It uses the fact that density
ρ equals Nm_p/V. (The mass of a sample is the product of the number of
particles and the mass of each.)

*At STP the density of hydrogen is 0.09 kg m^{-3}. Find the root mean
square velocity of the particles.* [122]

7.7 Some more questions

1. *Find these pressures.*
   (a) *At the 0.1 mm^2 tip of a drawing pin when 50 N is
       applied* [239]
   (b) *20 m below a lake surface* [334]
   (c) *2000 km high in Jupiter's atmosphere (mean density 500 kg m^{-3};
       g 27 m s^{-2})* [170]
2. (a) *5 m^3 of dry air at 100 kPa and 27 °C is compressed to 1 m^3.
       What is the new pressure?*
   (b) *The sample is next raised to 327 °C without change of volume.
       What does the pressure become?*
   (c) *The sample now expands to 3 m^3 without change of pressure.
   Find the final temperature.*
   (d) *Would your answers have been correct for damp air?* [500]
3. *In a factory argon is stored in a tank at 20 MPa and 27 °C. In an
   explosion and fire the tank volume is halved and the temperature rises
   to 227 °C. Will the tank burst if it is designed to withstand
   75 MPa?* [503]
4. *At one depth in the Sun the temperature is 10 MK; the pressure is
   4×10^{13} Pa. 1 m^3 of helium from that level is removed to STP. What
   is its new volume?* [302]
5. *5 kg of neon at 27 °C and 50 kPa has a density of 0.41 kg m^{-3}
   (a) What is the volume? (b) What would be the density at
   STP?* [431]
6. *Derive pV $=$ Nm$_p$v^2 as was done in section 7.7. Instead of using a
   1 m^3 box, take a volume V $=$ x $\times$ y $\times$ z.* [123]
7. *At STP, the density of radon is 10 kg m^{-3}. Find (a) its specific gas
   constant, [421] (b) the root mean square velocity of its
   particles.* [323]
8. *What is the root mean square velocity of radon particles at
   1092 K?* [424]
9. *The root mean square velocity of the particles of a gas sample is
   2 km s^{-1}. The gas is compressed to half its volume without change of
   pressure. Find the new root mean square velocity.* [252]

7.8 Objectives

Note: The first ten objectives are from Level 1.

When you have studied this chapter, you should be able to

(1) define density and solve simple problems relating to it;
(2) define pressure, state its unit, and calculate it using $p = F/A$;
(3) describe the pressure exerted by fluid to be the result of particle bombardment;
(4) state that the pressure at a point in a fluid is equal in all directions;
(5) state that fluid pressure does not depend on the shape of the container;
(6) state that fluid pressure acts at 90° to any contact surface;
(7) state that fluid pressure depends on the density and the depth;
(8) state that there is a pressure due to the atmosphere;
(9) state that barometers measure the pressure of the atmosphere;
(10) state the effect of 'heat' on the dimensions of solids, liquids and gases;
(11) define Boyle's law, and give an expression for it;
(12) define Charles' law, and give an expression for it;
(13) define the pressure law, and give an expression for it;
(14) convert temperatures between celsius and absolute values;
(15) define an ideal gas;
(16) solve problems involving changes of pressure, volume and temperature for ideal gases;
(17) state the gas laws for an ideal gas in the form $pV = nRT$,
(18) derive simply $pV = nRT$;
(19) relate the absolute temperature of a sample to the kinetic energy of its particles.

Chapter 8

Vapours

8.1 Evaporation

Vapours form a class of matter in the gas state. However they are far from ideal in how they behave. As many common gases at normal temperatures are in fact vapours, we need to find out about them.

Let us look again at how solids, liquids and gases differ from each other. (See Fig. 7.1)

One major point to emerge from Chapter 7 concerns the particle energies. We found that in a gas the particles do not all have the same energy. This is true, too, of the particles of solids and liquids. Some move more slowly than the mean speed; others move more quickly.

We can show this by a graph. The full line in Fig. 8.1 gives the range of particle energies in a sample at temperature T_1. The peak position relates to the mean value. We see that few particles have very low energy; most have energies close to the mean; very few have very high energy. In the same way the dashed curve shows the particle energy range at T_2 where T_2 is twice T_1. The mean energy in this case is twice as much as before. (See section 6.1 – **'temperature' measures the particles' mean energy**.) In the graph $N(W)$ is the number of particles of energy W.

The next point to make also comes from a statement in section 6.1.

The particles tend to attract each other.

At low energies the particles hold on to each other very tightly. So they form the rigid structure of a **solid** (Fig. 7.1 (a)).

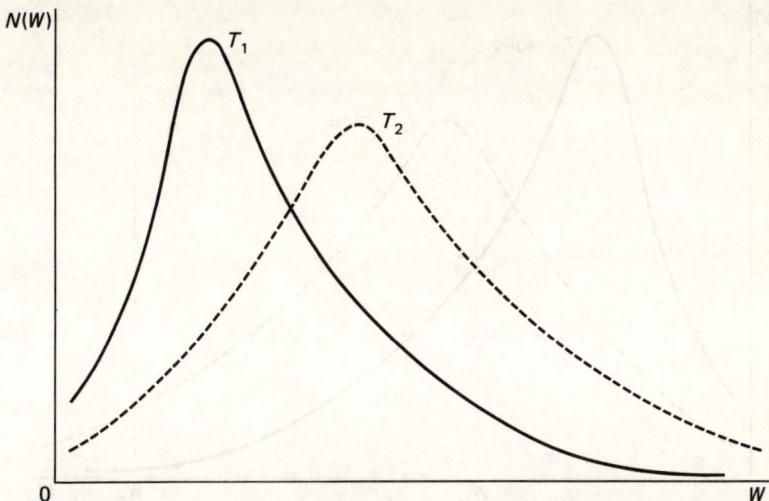

Fig. 8.1 The particles in a sample do not all have the same energy

At a certain temperature the particles have enough energy to break these bonds. Now as more energy is added, the particles escape from each other and flow – the sample becomes a **liquid** (Fig. 7.1(b)). Liquids are produced from solids by the change of state called **melting** (or *fusion*).

In a liquid the forces between the particles have much less effect. However, they are still strong enough to keep the particles together as a group even if each one is more free to move around. (We shall come back to this in Chapter 9.)

As yet more energy is given to the liquid, the temperature rises further. At a certain value the particles have enough energy to break quite free of the forces between them. More energy causes them to escape into the **gas** form of Fig. 7.1 (c). This change of state is **boiling**.

Melting and boiling involve the breaking of bonds. They require energy.

In the light of this, let us adapt the last figure, to give Fig. 8.2.

This time we have chosen values of T_1 and T_2 at which the sample is solid and liquid respectively. Also we have marked the energy required for melting – W_m – and the energy for boiling – W_b. These two figures relate to the melting and boiling temperatures T_m and T_b.

Some particles in solids and liquids have enough energy to escape into the gas state.

Check this with the graphs. Some particles in both samples have more energy than W_b.

If such high-energy particles are at the sample surface they may escape and form a gas. This process is **evaporation**.

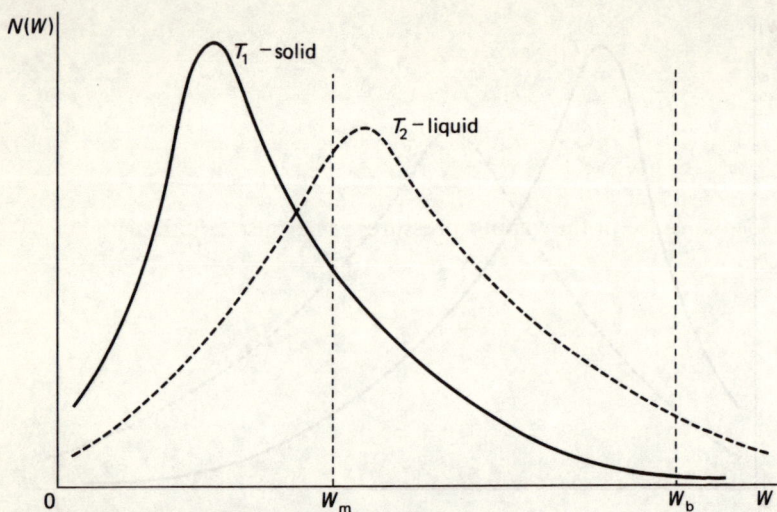

Fig. 8.2 The distribution of particle energies in a solid and in a liquid

However, only a very small number of particles can evaporate, unless T is close to T_b. This is because few particles have enough energy. So evaporation from a solid is rarely important. But it does happen – sometimes we can see snow evaporate, and if we can smell a solid, it clearly has a vapour. Evaporation from a liquid is often very important. Thus evaporation from lakes in hot countries can be a major problem.

In the rest of this chapter, all that we say about evaporation of liquids applies to solids as well.

Figure 7.1 should be changed to allow for all this – Fig. 8.3.

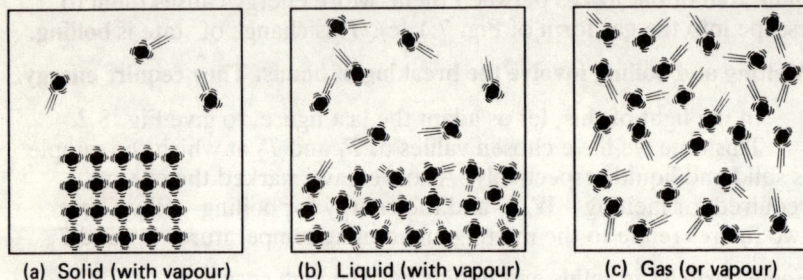

(a) Solid (with vapour)   (b) Liquid (with vapour)   (c) Gas (or vapour)

Fig. 8.3 A better view of the particles of matter

8.2 Saturated vapour pressure

For the time being we say that **a vapour is the result of evaporation**. Later we shall be more precise.

Because a vapour is in the gas state, it exerts a pressure on the walls of the box. This **vapour pressure** depends on the number of vapour particles in unit volume. To discuss this further, we start with an empty box.

As soon as we put a liquid in the bottom, evaporation starts.

In the space above the liquid, vapour particles will move around at high speeds. They collide with each other; they collide with the walls (this is the cause of the vapour pressure). They also collide with the liquid surface. Those that do this return to the liquid state – they **condense**.

As time passes, more and more liquid particles escape. The vapour pressure rises. As a result more and more vapour particles hit the liquid surface and condense.

At last the vapour will reach balance (**equilibrium**). Then, in one second, as many particles leave the liquid as return to it. The space is now **saturated** with vapour – it can hold no more.

A saturated vapour is in equilibrium with the liquid (or solid) state.

If the space contains less vapour than it could, it is **unsaturated**.

We have seen that the pressure of a vapour is called vapour pressure. In the same way, **the pressure of a saturated vapour is called the saturated vapour pressure** ('s.v.p.' for short).

Do saturated vapours obey the **ideal gas laws**?

Thus, if the volume of the space is halved, does s.v.p. double (Boyle's Law)? Or, if the absolute temperature doubles, does s.v.p. do the same (pressure law)?

The answer to these questions is NO!

It is not hard to test saturated vapours for gas law behaviour. Let us think what would happen.

In Fig. 8.4(a) a saturated vapour fills the space. Figure 8.4(b) shows the volume halved. The space now contains a vapour whose pressure is twice s.v.p. (That makes it a **supersaturated vapour**.) That means that twice as many vapour particles enter the liquid as liquid-particles leave it. Within a short time half the vapour condenses (returns to liquid form) – and the pressure equals s.v.p. again. (See Fig. 8.4(c).) Note that in *sudden* changes like this the condensation is

(a)

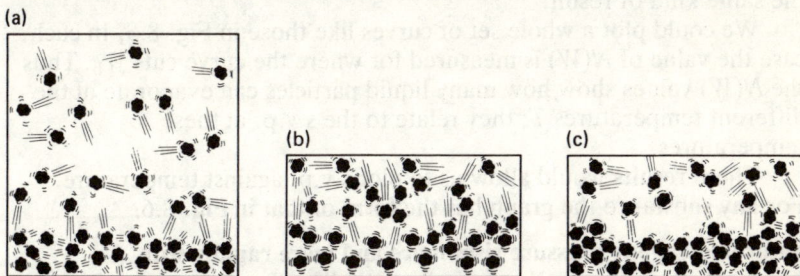

Fig. 8.4 A saturated vapour does not follow Boyle's Law

likely to take place on the walls – as **dew** – and in the space – as **fog** – as well as on the liquid surface. Fog and dew are forms of **precipitation. Precipitation is the condensation of supersaturated vapour.**

Saturated vapour pressure does not depend on volume – saturated vapours do not follow Boyle's law.

Now let us think about the pressure law. To test this we double the saturated vapour temperature; volume is kept constant. Does the s.v.p. double? We must recall some ideas from the last section.

Here are graphs like those of Fig. 8.1. T_2 is twice T_1. The dashed line W_b relates to the boiling temperature T_b of the substance.

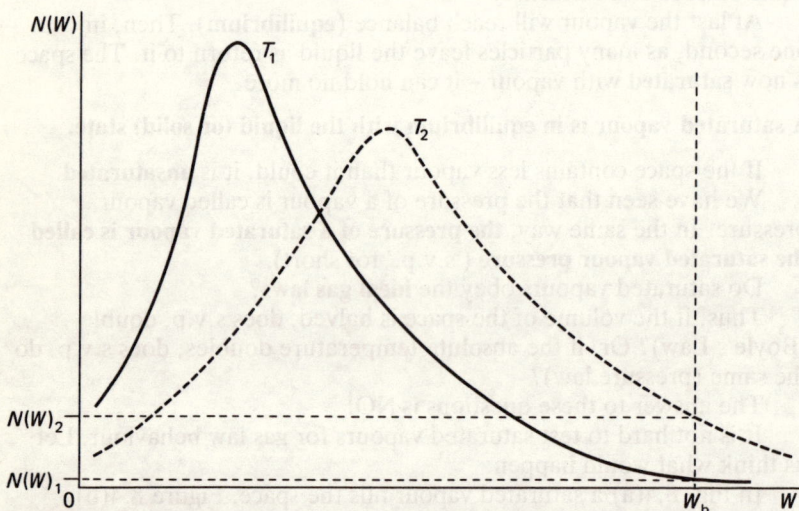

Fig. 8.5 The distribution of liquid particle energies at two temperatures

Clearly, at T_2 far more than twice as many liquid particles evaporate than at T_1. The numbers in the two cases are $N(W)_2$ and $N(W)_1$. Testing any saturated vapour over any temperature range gives the same kind of result.

We could plot a whole set of curves like those in Fig. 8.5. In each case the value of $N(W)$ is measured for where the curve cuts W_b. Thus the $N(W)$ values show how many liquid particles can evaporate at the different temperatures T; they relate to the s.v.p. at these temperatures.

These results would allow us to plot s.v.p. against temperature. For any substance the graph has the form of that in Fig. 8.6.

Saturated vapour pressure rises more and more rapidly with temperature – saturated vapours do not follow the pressure law.

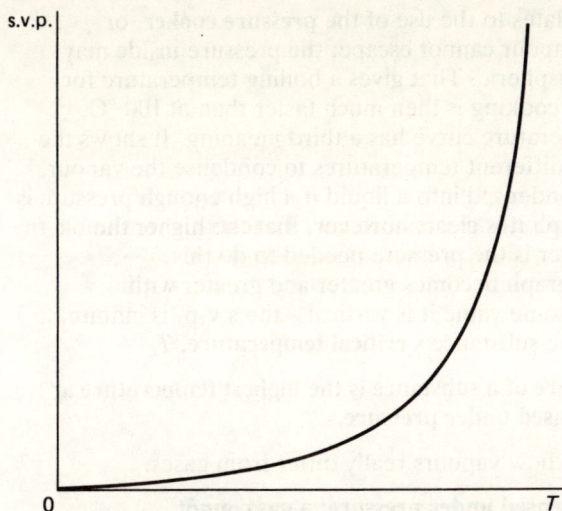

Fig. 8.6 A saturated vapour pressure–temperature curve

The box gives some basic facts about s.v.p.; we have discussed the first two, but the others are true as well.

(i) The saturated vapour pressure of a substance depends only on temperature.
(ii) It does *not* depend on volume.
(iii) It does *not* depend on what other gas(es) may be present.
(iv) It does *not* depend on the pressure of the other gas(es) that may be present.

What does this mean? A saturated vapour in a space is in equilibrium with its liquid (or solid) form. The s.v.p. will be the same whether the space is 1 mm^3 or 1 km^3. It will be the same whether the space contains no other gas or it contains air at 10 MPa. The saturated vapour pressure of a substance depends *only* on temperature.

It is also useful to know that **a liquid boils when its s.v.p. equals the outside pressure**. The same is true when a solid sublimes. If the s.v.p. is below the outside pressure, not many liquid particles can escape; when s.v.p. equals the outside pressure, all can pass into the gas state.

Thus the graph of Fig. 8.6 also shows how boiling temperature depends on outside pressure.

The concept relates to the well known fact that water boils further below 100 °C as altitude increases. (That is why British mountain climbers are said to have trouble making good hot tea!)

The graph also relates to the use of the **pressure cooker**, or **autoclave**. Here the vapour cannot escape; the pressure inside may then reach twice atmospheric. That gives a boiling temperature for water of over 120 °C; cooking is then much faster than at 100 °C.

The s.v.p. – temperature curve has a third meaning. It shows the pressures required at different temperatures to condense the vapour. Any vapour can be condensed into a liquid if a high enough pressure is applied. From the graph it is clear, however, that the higher the temperature, the higher is the pressure needed to do this.

The slope of the graph becomes greater and greater with temperature rise. At some value it is vertical – the s.v.p. is infinite. This temperature is the substance's **critical temperature**, T_c.

The critical temperature of a substance is the highest temperature at which it can be condensed under pressure.

Now we can state how vapours really differ from gases.

A vapour can be condensed under pressure; a gas cannot.

In other words, a substance above the critical temperature can only be a 'gas'; it is a 'vapour' below it unless it is condensed.

So the curve in Fig. 8.6 can be drawn with three sets of axes. They are s.v.p. – absolute temperature; the pressure at which the substance will boil against temperature; and the pressure needed to condense the substance at different temperatures.

The graph in Fig. 8.7 shows how the s.v.p. of water varies between 0 °C and 200 °C. Use the graph to estimate (a) the s.v.p. of water at (i) 27 °C and (ii) 127 °C; (b) the boiling temperature at (i) 50 kPa and (ii) 150 kPa; (c) the condensation pressure at (i) 50 °C and (ii) 200 °C; (d) the critical temperature. [297]

8.3 Evaporative cooling

Evaporation is the escape of high energy particles from a surface.

It is clear from this that evaporating particles have higher energies than the mean. We also see this from Fig. 8.5 – it is the particles to the right of the dashed line W_b that can escape.

If high-energy particles leave a sample, its mean energy falls. In other words, **evaporation causes cooling**. (If the sample can gain enough energy from elsewhere, of course this would not occur.)

The effect is called **evaporative cooling**; it is well known. Because of it, a damp cloth is cooler than the dry one next to it. Many people put damp cloths over bottles of milk and packs of butter to keep them cool in hot weather. Damp cloths are also used to reduce fever.

For the same reason, the main function of **sweating** is to cool the body. The blood temperature may rise too high because of hard work,

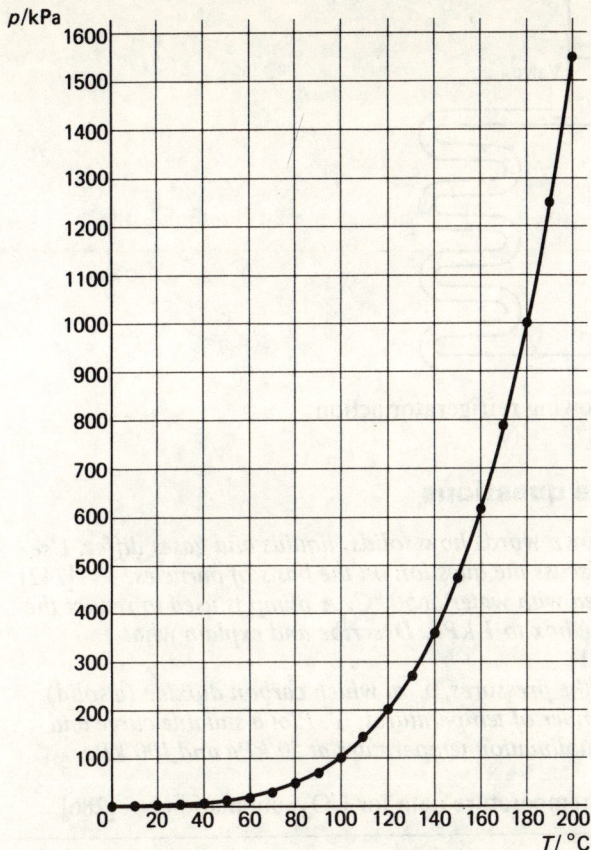

Fig. 8.7 Part of the s.v.p. – temperature curve for water

hot weather or fever. Then the sweat glands in the skin produce a liquid. As this evaporates, the skin – and thus the blood – is cooled.

We use evaporative cooling in **refrigeration**. This includes the action of freezers, cold-stores and air-conditioners. Figure 8.8 outlines how such a system works.

A **volatile** (easily evaporated) fluid is used. Often this is a special substance called a *freon*. It is pumped around the circuit. As it passes through the pump, it is a vapour. The high pressure makes it condense; it also makes it warmer. The excess energy is released to the outside as the liquid moves through the coil C to the tank T. The liquid then passes through the valve V to the low-pressure side of the circuit. This causes it to start evaporating, and also cools it. For both reasons the fluid takes energy from outside the freezing coil F.

In a refrigerator, F is in the freezing space and C is outside, at the back.

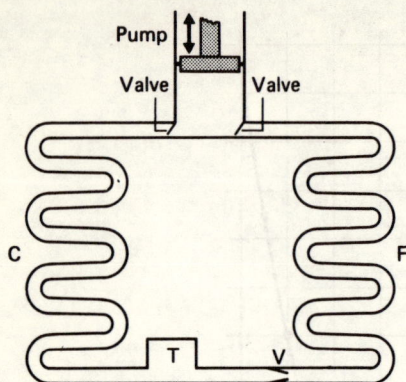

Fig. 8.8 Scheme showing refrigerator action

8.4 Some more questions

1. *Explain in your own words how solids, liquids and gases differ. Use diagrams, and discuss the question on the basis of particles.* [242]
2. *A box is part filled with water at 50 °C. A pump is used to reduce the air pressure in the box to 1 kPa. Describe and explain what happens.* [341]
3. *Table 8.1 shows the pressures, p, at which carbon dioxide (a solid) sublimes at a number of temperatures, T. Plot a suitable curve and from it find the sublimation temperature at 50 kPa and 100 kPa.*

Table 8.1 Pressure/temperature data for CO_2 sublimation [286]

$T/$ °C	−120	−115	−110	−105	−100	−95	−90	−85	−80
$p/$kPa	1.2	2.3	4.5	7.8	13.6	21.5	37.0	66.1	85.0

[286]

4. *Does boiling cause cooling?* [124]

8.5 Objectives

When you have studied this chapter, you should be able to

(1) describe evaporation in terms of the kinetic theory;
(2) relate evaporation to other state changes;
(3) define vapour pressure;
(4) define and explain saturated vapour pressure;
(5) discuss why a saturated vapour does not behave as an ideal gas;

(6) describe how saturated vapour pressure relates to temperature;
(7) explain the drop in temperature occurring with evaporation in terms of particles;
(8) discuss three interpretations of the s.v.p. – temperature curve;
(9) define critical temperature and distinguish between a vapour and a gas;
(10) draw and use s.v.p. – temperature curves.

Chapter 9

The states of matter

9.1 The effects of thermal energy

Thermal energy is energy supplied to the particles of matter or supplied by them. Thermal energy is sometimes called 'heat'; do not confuse it with temperature. Themal energy is energy moving to or from matter particles; temperature is a measure of the energy of the particles. In daily life we mustn't confuse the two; the same is true in physics. Temperature change (and the change of size that goes with it) is one of two possible effects of thermal energy on matter; the other, as we have seen, is change of state.

While working through the last few chapters, you should have learned how to relate each of these changes. This must be done (a) in terms of any sample, and (b) in terms of the particles of the sample.

In this section we check these ideas and others from work done at Level 1 or 'O' Level.

Thermometers are used to measure temperature.

Each type of thermometer is based on some physical property of matter; this depends in a known way on temperature.

The physical property used with the **mercury in glass thermometer** (Fig. 9.1) is the length of a liquid column.

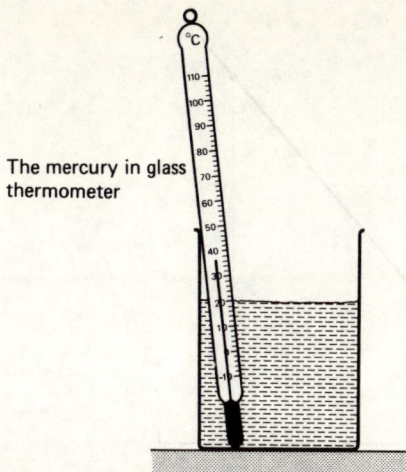

Fig. 9.1

What temperature does the thermometer in Fig. 9.1(a) show? [125]

All types of thermometer are marked by using certain **fixed temperatures** (or *fixed points*). Some of these were listed in section 7.3.

Fixed temperatures are easy to provide in practice; their values are agreed by all. The two most important to us are T_i, **the ice temperature** (0 °C, 273 K), and T_s, the **steam temperature** (100 °C, 373 K).

The ice temperature is the temperature of pure melting ice in equilibrium with liquid water at standard air pressure.

The steam temperature is the temperature of pure boiling water in equilibrium with water vapour at standard air pressure.

Why must the pressure be set at a certain value? [348]

So the thermometric property is measured at two fixed temperatures. If we know how the property varies with temperature, we can then draw a graph (a 'calibration' curve). Figure 9.2 shows the form of the graph we might have with the thermometer discussed above.

In use the thermometer's physical property is measured at the unknown temperature; the curve is then used to find what the temperature is. Often, of course, temperature readings are marked straight on to the thermometer.

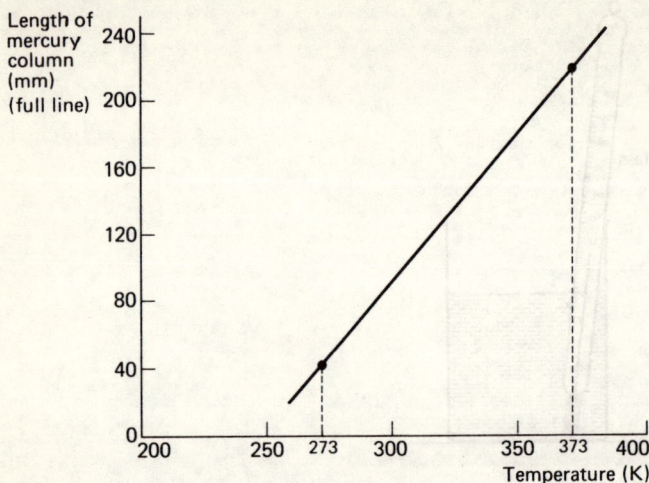

Fig. 9.2 Typical thermometric graph

Use the graph in Fig. 9.2 to find the temperatures when the length of the mercury column is (i) 30 mm (ii) 240 mm.   [220]

What happens if the same amount of energy is given to different samples?

We find that the resulting temperature change depends on (a) the substance used, and (b) the sample mass. The **specific thermal capacity** *c* of the substance is needed here. (This is sometimes called *specific heat*.)

The specific thermal capacity of a substance is the energy needed to raise unit mass by one degree.

$$\text{Specific thermal capacity} \quad = \quad \frac{\text{energy}}{\text{mass} \times \text{temperature change}}$$

$c = W/(m \times \triangle T)$ Unit: joule per kilogram per kelvin, J kg^{-1} K^{-1}
The equation is often used in the form $W = mc \triangle T$.

A Bunsen burner gives 1 kJ per minute. It is used to warm 250 g of water in a beaker. A quarter of the energy from the flame enters the water. After three minutes the water temperature has risen by 43 °C. Find the specific thermal capacity of the water.

$P = 10^3$ J/min t = 3×60s $\eta = 0.25$
$m_w = 0.25$ kg $\triangle T_w = 43$ °C(43 K) $C_w = ?$
$W_w = \eta\ Pt = 0.25 \times 10^3 \times 3 \times 60$J

$$W_w = m_m c_w \triangle T_w \rightarrow$$
$$c_w = W_w/m_w \triangle T_w = 0.25 \times 10^3 \times 3 \times 60/(0.25 \times 43) \text{ J kg}^{-1}\text{ K}^{-1}$$
$$\Omega \ 4200\text{J kg}^{-1}\text{ K}^{-1}$$

Note: You should learn an approximation to the value given by the last question.

A 10 kg block of copper cools from 500 °C to 50 °C. The specific thermal capacity is 420 J kg^{-1} K^{-1}. How much energy is released? [238]

The next change to look at is the 'change of size' that follows 'change of temperature'. This is **thermal expansion**.

All substances change size when their temperatures change. The expansion of solids and liquids is much less than that of gases.

We discussed how gases behave in Chapter 7. In section 7.3 we noted that **most substances expand with temperature rise**. The common

(a) The car lamp flasher unit *(from data supplied by Lucas Ltd.)*

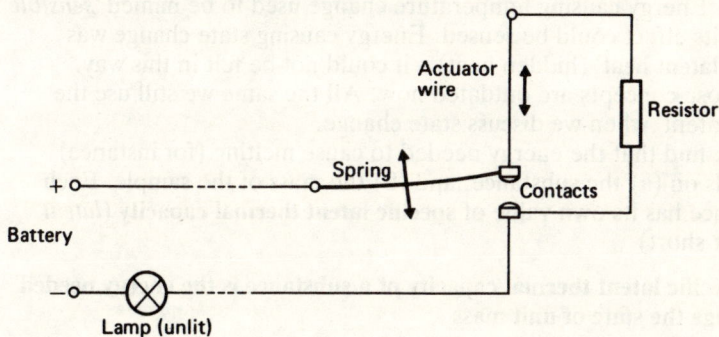

(b) A thermostat

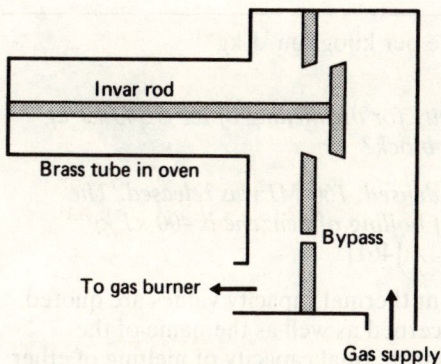

Fig. 9.3 Expansion in use

one that does not is **water** between 0 °C and 4 °C – in this range it contracts (its expansion is negative).

We do not need to know much more about expansion here. It exists; it causes problems in design which we must avoid; it has a number of uses.

For instance, the design of structures like roads, bridges, railway lines and large buildings must allow for thermal stresses and strains. The same is true for the design of mechanical clocks and watches.

Two **uses of expansion** are given in Fig. 9.3. You should explain how each acts.

9.2 'Latent heat'

Energy added to matter can have one of two results. The first is change of temperature; we have just dealt with this. The second is change of state.

The former can be observed with a thermometer; the latter cannot. Energy causing temperature change used to be named '*sensible heat*' – its effect could be sensed. Energy causing state change was called '**latent heat**' (hidden heat) – it could not be felt in this way.

Those concepts are outdated now. All the same we still use the word 'latent' when we discuss state change.

We find that the energy needed to cause melting (for instance) depends on (a) the substance, and (b) the mass of the sample. Each substance has its own value of **specific latent thermal capacity** (*latent heat* for short).

The specific latent thermal capacity of a substance is the energy needed to change the state of unit mass.

Specific latent thermal capacity = energy/mass

$$L = W/m \quad \text{Unit: joule per kilogram, J kg}^{-1}$$
This gives $W = mL$.

The specific latent thermal capacity for the melting of ice is 340 kJ kg^{-1}. What energy would melt a 500 g block?

When some benzene vapour condensed, 160 MJ was released. The specific latent thermal capacity of boiling of benzene is 400 kJ kg^{-1}. How much vapour condensed?   [461]

Note that when specific latent thermal capacity values are quoted, we include the state change concerned as well as the name of the substance. Thus the specific latent thermal capacity of melting of ether is 102 kJ kg^{-1}; the specific latent thermal capacity of boiling of ether is 350 kJ kg^{-1}. For boiling the value is always higher than for melting.

Specific latent thermal capacity values also depend on **temperature**. If no temperature is given, assume that it is the standard state change temperature – that for pure substance at standard air pressure.

Here is a list of state changes.

Table 9.1 State changes and their names

State change	Name	Temperature
Solid → liquid	melting (fusion)	melting temperature T_m
Liquid → solid	freezing (solidification)	freezing temperature $T_f = T_m$
Solid → gas	sublimation	sublimation temperature T_s
Gas → solid	freezing (solidification)	freezing temperature $T_f = T_s$
Liquid → gas	boiling (vaporisation)	boiling temperature T_b
Gas → liquid	condensation (liquefaction)	condensation temperature $T_c = T_b$

Each state change takes place at a certain temperature under standard conditions. A certain value of specific latent thermal capacity then relates to it.

However, in the last chapter we saw that any liquid or solid evaporates to some extent at any temperature. The reverse is true too – any vapour can condense or freeze to some extent at any temperature. (Recall that a gas is a 'vapour' only below the critical temperature – section 8.2.)

As evaporation is a state change, energy is involved. That energy is the specific latent thermal capacity for the change concerned at the temperature in question. A substance's specific latent thermal capacity for evaporation falls as temperature rises. This fall relates to the s.v.p./temperature curves discussed in section 8.2.

The next question uses these ideas. It should also help us to recall the concept of **evaporative cooling** (section 8.3).

200 kg of water at 20 °C lies on a flat roof. One per cent evaporates. How much will the temperature of the pool fall if no energy comes from outside? The specific latent thermal capacity of water at 20 °C is 2.5 MJ kg^{-1}; the specific thermal capacity of water is 4200 J kg^{-1} K^{-1}. [313]

50 g of ether is evaporated (a) at 0 °C (b) at 100 °C. The two values of specific latent thermal capacity are 395 kJ kg^{-1} and 325 kJ kg^{-1}. Find the energy released in each case. [128]

9.3 Surface tension

The effect of surface tension is often described as being like that of an elastic skin on a liquid surface. A common trick based on it is shown in Fig. 9.4(a).

In fact there is *no* skin on a liquid surface – but the effect does exist.

What is surface tension?

To support that needle, the water must apply an upward force on it. Look at Fig. 9.4(b). Other effects of surface tension are seen in Fig. 9.5; all involve forces.

Once more we must think about matter particles and how they behave. In fact surface tension strengthens the view that matter consists of particles – it is hard to explain any other way.

Here is the **kinetic theory** statement that we use.

The particles of matter attract each other.

In Fig. 9.6(a) a particle is shown inside a volume of liquid. To make the point more clearly, we ignore its motion and weight. The arrows give, in strength and direction, the forces on the particle from

(a) Floating a needle on a water surface

(b) A sectional view of the needle and the forces acting on it

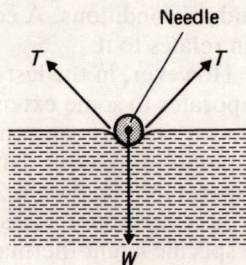

Tissue paper used to support needle at first

Needle

Fig. 9.4 Surface tension

(a) At the edge of a liquid surface, the surface is pulled into a **meniscus**

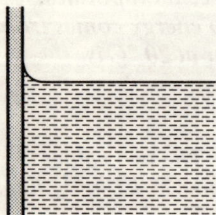

(b) Liquid **drops** tend to be pulled into spheres

(c) Liquids rise in narrow (**capillary**) tubing

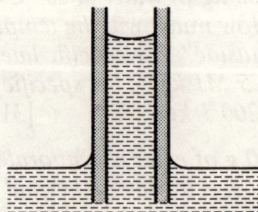

Fig. 9.5 Other effects of surface tension

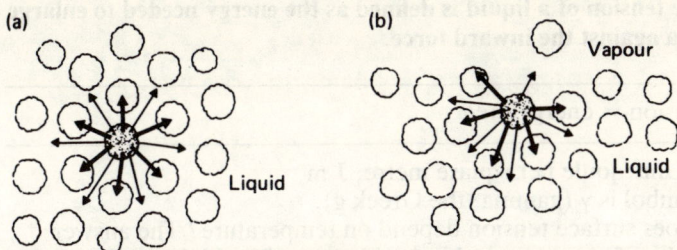

Fig. 9.6 The forces on a particle in a liquid

its neighbours. Clearly, on balance the net attractive force is zero, since all the forces cancel each other out.

In Fig. 9.6(b) a particle is shown near a liquid surface. This time we see that there is a net attractive force from the neighbours – the resultant acts into the liquid.

A particle of the liquid at its surface is pulled inward.

This has the effect that the liquid sample tries to keep the surface area as small as possible. A drop has the shape of a sphere. A needle on the surface extends the surface, so is pushed up.

We can discuss surface tension in terms of potential energy in the liquid surface. This is because the surface can apply forces and do work. Refer again to Fig. 9.5. Two common experiments that show this in more detail are shown in Fig. 9.7.

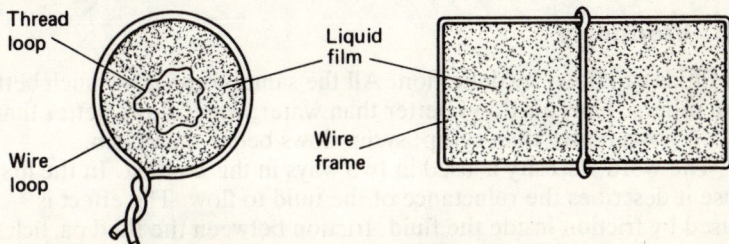

What happens when the film inside the thread is broken?
Work is done; the thread is moved.

What happens when the right-hand film is broken?
Work is done; the slider is moved.

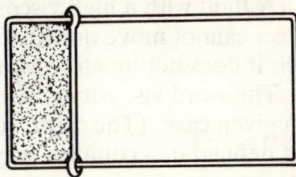

Fig. 9.7 Energy is stored in a liquid form – it can apply force and do work

The surface tension of a liquid is defined as the energy needed to enlarge surface area against the inward force.

Surface tension = energy/area

$\gamma = W/A$. Unit: joule per square metre, J m^{-2}
   The symbol is γ (gamma, the Greek g).
   How does surface tension depend on temperature? The answer should be clear from what we have said above. The higher the temperature, the more the particles tend to break free of each other. Therefore –

Surface tension falls with temperature rise.

   This is because the particles move apart as the temperature rises. (This is **thermal expansion** of course.) Thus they do not hold on to each other so tightly.
   We would expect this. We know from section 8.2 that there is much more evaporation at higher temperatures. And in section 9.2 we found that a liquid's **specific latent thermal capacity** falls with temperature rise.
   In fact, surface tension relates closely to specific latent thermal capacity for evaporation. Both fall with temperature in the same way; both are zero at the **critical temperature** of the substance. See question 5 in section 9.5.

9.4 Viscosity

All fluids can flow, by definition. All the same, some flow much better than others. Thus air flows better than water; water flows better than oil; oil flows better than syrup; syrup flows better than pitch.
   The word **viscosity** is used in two ways in this context. In the first sense it describes the reluctance of the fluid to flow. The effect is caused by friction inside the fluid, friction between the fluid particles. In this sense, viscosity is often now called **fluid friction**.

The viscosity (fluid friction) of a fluid is its reluctance to flow.

   A fluid with a high viscosity does not flow very quickly; also an object cannot move quickly through it. A fluid of low friction flows well; it does not much resist motion through it.
   The word viscosity is also used as a *measure* of reluctance to flow in a given case. (The term 'coefficient of viscosity' still often appears.) It is defined in a complex way; the details do not matter at this stage.

The (coefficient of) viscosity of a fluid measures its reluctance to flow.

   The symbol is η (eta, Greek e); the unit is the pascal second, Pa's.

Fluid friction affects both rate of flow and resistance to motion through the fluid, as we saw above. In experiments to measure viscosity we use either effect.

Thus we can measure the **terminal speed** of an object falling freely through the fluid.

In section 1.2 we discussed the acceleration of **free fall**; we made the point that for this to be constant, there must be no friction. In fact, as a result of friction, objects fall through a fluid with a smaller acceleration than we would expect. Also they go no faster than a certain speed. This is the 'terminal speed' of the object in the fluid.

The terminal speed of a person falling from height through air is around 100 m s^{-1}. A **parachute** is far more subject to air friction – there is far more 'drag'. With a parachute, a falling person's terminal speed is about 8 m s^{-1}. Thus fluid friction, viscosity, may be of use.

If there were no air friction, from what height would a person fall to reach (a) 100 m s^{-1}, (b) 8 m s^{-1}?   [380]

A person falls from 10 km. At what speed would she reach the ground if there were no air friction?   [354]

Thus, although the viscosity of air is very low, its effects cannot be ignored. How low the viscosity of air is, compared with that of other fluids, is shown in Table 9.2.

Table 9.2 Sample viscosities

Substance	Temperature/° C	η/Pa s
Water	0	0.18
Water	100	0.03
Ether	0	0.03
Ether	100	0.01
Syrup	15	400
Syrup	25	140
Air	0	1.7×10^{-3}
Air	100	2.2×10^{-3}
Neon	0	3.0×10^{-3}
Neon	100	3.7×10^{-3}

We can see these facts about fluid friction.
 (i) Values for gases are much lower than for liquids.
(ii) Liquid viscosity falls quickly with increase of temperature. (This is well known for syrup and engine oil.)
(iii) Gas viscosity rises with increase of temperature.

All this can be explained best on the basis of kinetic theory. Fluid friction effects are caused by friction between the moving particles of the fluid.

9.5 Some more questions

1. *Describe other types of thermometer than that noted in this chapter. Explain the physical properties they use; how do these vary with temperature?* [244]
2. *A 25 kg copper tank (specific thermal capacity 400 J kg^{-1} K^{-1}) contains 100 kg of water at 15 °C. It is warmed by a 3 kW immersion 'heater'. If there are no energy leaks, how long is needed to raise the temperature to 55 °C?* [336]
3. *See Fig. 9.8. All the ether evaporates. How much ice forms round the can? The specific latent thermal capacities concerned are – ether: 400 kJ kg^{-1}; water: 320 kJ kg^{-1}.* [474]

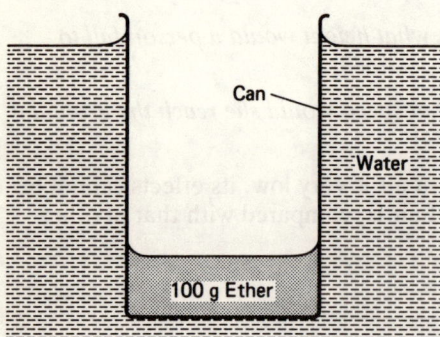

Fig. 9.8

4. *For boiling, the specific thermal capacity value is always higher than for melting (section 9.2). Why?* [324]
5. *Table 9.3 gives these data for water between 0 °C and 100 °C: specific latent thermal capacity of evaporation, L: surface tension, γ. Using the same T-axis, devise an L-axis (vertical left) and a γ-axis (vertical right) and plot the data. Compare how the two measures vary with temperature* T. [253]
6. *Both the surface tension and the specific latent thermal capacity of evaporation of a liquid are zero at the critical temperature (section 9.3). Why?* [414]
7. *Explain the facts about fluid friction (section 9.4) on the basis of particles.* [455]

Table 9.3

$T/°\text{ C}$	$L/\text{J kg}^{-1}$	$\gamma/\text{mJ m}^{-2}$
0	2.50	75.7
10	2.48	74.2
20	2.45	72.8
30	2.43	71.2
40	2.41	69.6
50	2.38	67.9
60	2.36	66.2
70	2.33	64.4
80	2.31	62.6
90	2.28	60.7
100	2.26	58.8

9.6 Objectives

Note: The first eight objectives are from Level 1.
When you have studied this chapter, you should be able to
(1) differentiate between temperature and 'heat';
(2) define fixed temperatures and give two examples;
(3) describe how to measure temperature;
(4) discuss the effect of thermal energy on the physical dimensions of matter samples;
(5) give examples of its practical applications and design implications;
(6) differentiate between 'sensible heat' and 'latent heat';
(7) describe evaporation in terms of the kinetic model;
(8) define and use specific latent thermal capacity;
(9) calculate energy losses due to evaporation using specific latent thermal capacity values;
(10) describe how surface tension effects can be explained by forces between liquid particles.
(11) recognise a relation between latent thermal capacity and surface tension data;
(12) describe (coefficient of) viscosity as indicating reluctance to flow, and its measurement;
(13) discuss the relation between terminal speed and fluid friction;
(14) state the variation of liquid and gas viscosities with temperature.

Part 3

Waves and vibrations: the transfer of energy

Chapter 10

What are waves?

10.1 Waves and energy

There are two main ways in which energy can pass from one place to another.

Firstly, **particles** can carry it. Thus electrons carry electrical energy; particles of matter carry thermal energy by convection and evaporation. You should have met these ideas before.

On the other hand, energy can travel by some kind of **vibration**. This we call a **wave**.

In this chapter we discuss a few types of wave. All behave in the same kind of way; all carry energy through a **medium**. The word 'medium' describes the space – matter or a vacuum – through which energy travels.

Many types of waves are known or thought to exist. Here are some.

Liquid surface waves – for instance, the waves made when a stone drops into a pool.

Sound waves – as produced by the vibrating prongs of a tuning fork. Sound waves are often called **pressure waves**.

Light waves – we know that light carries energy. Plants use light to make food. Photocells change light energy into electricity; this can move a needle or power a radio. In the next chapter we meet proof that light travels in waves.

Here we shall mainly discuss these three wave forms. However, there are others.

Energy passes in the form of a wave if we can draw a graph like that in Fig. 10.1. Here y is the physical property that varies in value. In the case of sound, y is the pressure in the medium. For liquid surface waves, y relates to the height of the surface. With electromagnetic waves, like light and radar, y is the strength of the electric (or magnetic) field.

For any wave, a number of crucial terms are used. Some are shown in Fig. 10.1. They relate to complete **cycles** of vibration. A cycle is a full to and fro change, coming back to the starting point. Figure 10.1 shows three cycles.

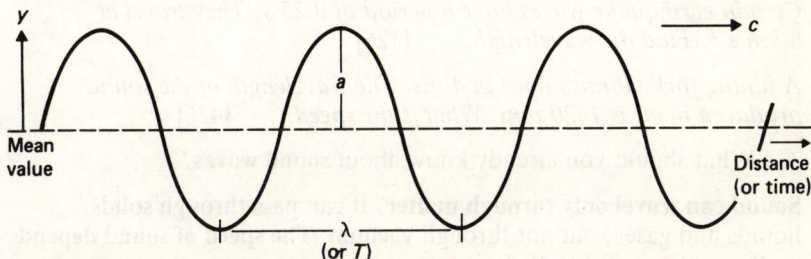

Fig. 10.1 A wave motion

The wavelength λ (lamda) of a wave motion is the distance covered by one cycle. The unit is the metre (m).

The period T of a wave motion is the time taken by one cycle. Unit – second (s).

The frequency ν of a wave motion is the number of cycles in unit time. Unit – hertz (Hz).

The speed c of a wave motion is the distance moved in unit time. Unit – metre per second (m s^{-1}).

The amplitude a of a wave motion is the difference between the peak value and mean value of y. The unit is the unit of y.

Amplitude relates to the rate of energy transfer by the wave. The amplitude of a sound wave involves its loudness; the amplitude of a light wave concerns brightness, and so on. In other words, the larger the amplitude, the more **intense** the wave.

The other four wave quantities are connected by the following formulas.

Period = 1/frequency

Speed = frequency × wavelength = wavelength/period

The equations in the second block are forms of **the basic wave equation**. All apply to *any* wave.

The distance between peaks of waves on a pond is 0.5 m. Twenty waves reach the bank in four seconds. Find the period. What is the speed?

$\lambda = 0.5$ m $\nu = 20/4$ Hz $= 5$ Hz $T = ?$ $c = ?$
$T = 1/\nu = 1/5$ s $= 0.2$ s
$c = \nu\lambda = 5 \times 0.5$ m s^{-1} $= 2.5$ m s^{-1}

BBC Radio 2 transmits on 1500 m. The speed of radio waves is 300 000 km s^{-1}. Find the frequency and period. [204]

Certain earthquake waves have a period of 0.25 s. They travel at 8 km s^{-1}. Find the wavelength. [126]

A tuning fork vibrates once in 4 ms. The wavelength of the sound produced in air is 1320 mm. What is the speed? [426]

What should you already know about **sound waves**?

Sound can travel only through matter. It can pass through solids, liquids and gases, but not through vacuum. **The speed of sound depends on the medium.** Table 10.1 gives some data.

Table 10.1 Speed of sound in different media

Medium	Speed of sound c/m s^{-1}
Dry air at STP	330
Water at 25 °C	1500
Steel	6000

The frequency of sound depends on the size of the source. Large objects produce low-frequency sound – a war drum, a double bass, a tuba, a man. Small objects produce high frequency sound – a hand bell, a violin, a penny whistle, a bat.

   Musical instruments provide many useful examples of sound sources. They make sound by vibration of:

 (i) Stretched strings: the string instruments, harp, piano.
 (ii) Columns of air: the wind instruments.
(iii) Solid objects or surfaces: bells, triangle, drum skin.

The human ear is sensitive only to a certain sound frequency range. This is *about* 16 Hz to 16 kHz. Sounds of lower frequencies form **infrasound**; sounds of higher frequencies are **ultrasound**.

The lowest note of a violin has a frequency of about 200 Hz. What is its wavelength in air? (See Table 10.1.) [138]

Some whales squeak notes of wavelength 30 mm in water. What is the frequency? (See Table 10.1.) [129]

10.2 Longitudinal and transverse waves

We have seen that there are many kinds of wave. All can be described in terms of a graph like that in Fig. 10.1; all have wavelengths and frequencies related to wave speed by $c = \nu \lambda$. All behave in the same way.

One way to describe a wave concerns the direction of its vibration. This can be either *along* the direction of travel, or *across* it (or both).

In a 'longitudinal' wave the vibrations are along the direction of travel.
In a 'transverse' wave the vibrations are across the direction of travel.

Fig. 10.2 Longitudinal and transverse waves in a length of spring

Figure 10.2(a) shows a long spring, resting on the ground; it is fixed to a wall at one end. On one of the links near the wall is a piece of paper.

Figure 10.2(b) shows how we can vibrate the free end back and forth *along* the length of the spring. All the links – and the piece of paper – move back and forth the same way. This models a **longitudinal wave**. Sound travels like this – the matter particles vibrate back and forth along the direction of travel.

Figure 10.2(c) shows how we can vibrate the free end to and fro *across* the length of the spring. The links move the same way. This is a **transverse wave**. Light travels like this.

Note that liquid surface waves are a mixture of transverse and longitudinal components. At the surface the particles move in circles; further down their paths are ellipses (or smaller circles in deep water). Figure 10.3 shows this.

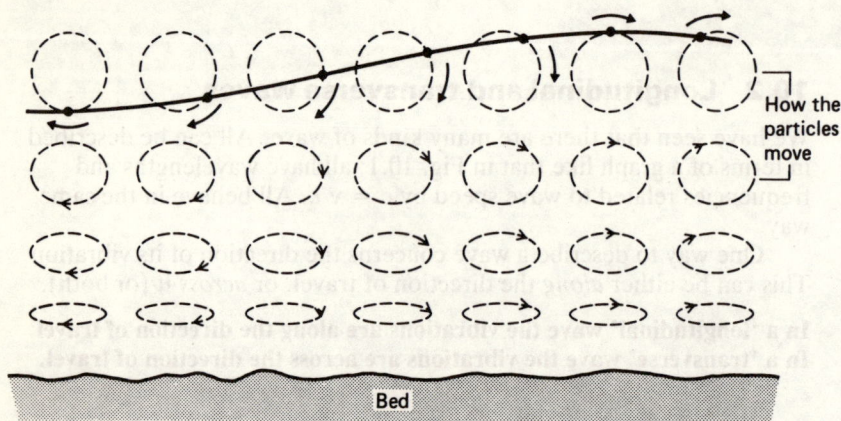

How the particles move

Bed

Fig. 10.3 Liquid surface wave motion is longitudinal and transverse

10.3 Polarisation

In a longitudinal wave the vibrations are along the direction of travel. In a transverse wave the vibrations are across the direction of travel.

We showed these two types of wave with a long spring. (See Fig. 10.2.) Sound and related waves are longitudinal; light and its relations are transverse waves.

A big difference between transverse and longitudinal waves is the number of possible directions of vibration. Longitudinal waves can vibrate one way only – to and fro the way the wave is going. However, transverse waves can vibrate in many ways across the direction of travel.

Thus if a transverse wave comes straight up out of the ground, it could vibrate north/south, or east/west, or northeast/southwest, or ESE/WNW, and so on. A longitudinal wave in this case could vibrate only up and down (along the travel direction).

Let us look at the vibrating spring again.

What would be the difference if the spring passed through a gap in a fence?

The fence would have no effect on longitudinal waves – energy could still move along the spring. Figure 10.4(a) shows this (from above).

But only up and down transverse waves could pass through the fence. To and fro (sideways) transverse waves could not pass at all. Transverse waves in other directions could not pass very well. (See Fig. 10.4(b).)

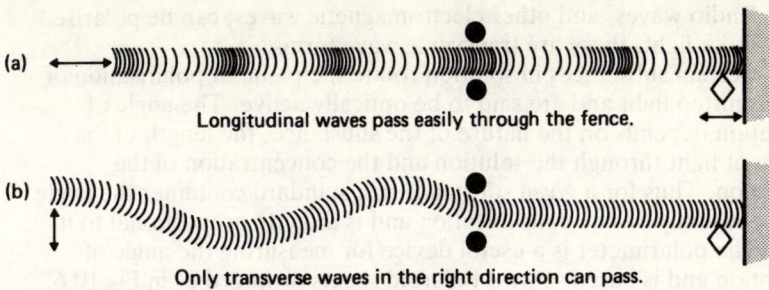

(a)

Longitudinal waves pass easily through the fence.

(b)

Only transverse waves in the right direction can pass.

Fig. 10.4 Polarisation of waves in a spring

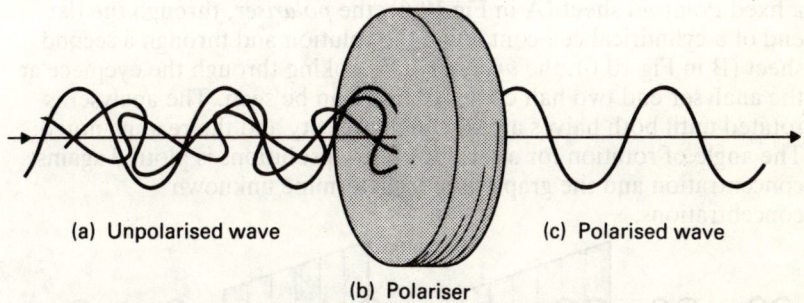

(a) Unpolarised wave

(b) Polariser

(c) Polarised wave

Fig. 10.5 Polarising an unpolarised wave

That is a useful model of '**polarisation**'. Polarisation is the passage of vibration in one direction but not in others. Only transverse waves can be polarised. This is because transverse waves can pass by vibration in any direction *across* the wave direction. (See Fig. 10.5(a).) Certain media will pass transverse waves only if they vibrate in a certain direction. Such media polarise incoming unpolarised waves. A **polariser** is shown in Fig. 10.5(b) – it passes only vertical vibrations. Its output (Fig. 10.5(c)) is vertically polarised.

Polarisation is the transmission of transverse vibration in a certain direction but not in others.

We have said that light is a transverse wave form. Therefore it can be polarised.

Polaroid is a substance that can polarise light. It can be used as shown in Fig. 10.6. Here we see that 'parallel' Polaroid sheets transmit (polarised) light; 'crossed' Polaroid sheets pass no light. Partly 'crossed' Polaroid sheets allow some light to pass, the amount being dependent on the angle of 'crossing'. Try this with two 'lenses' from Polaroid sunglasses.

This test shows that light can be polarised. That proves that light is a transverse wave form. Longitudinal waves, like sound, cannot be polarised; nor can streams of particles (not in this sense anyway).

106

Radio waves, and other electromagnetic waves, can be polarised too. Like light, these are transverse wave forms.

Certain substances in solution rotate the plane of polarisation of transmitted light and are said to be optically active. The angle of rotation depends on the nature of the substance, the length of the path of light through the solution and the concentration of the solution. Thus for a *given* substance in a standard container the angle depends only on the concentration and is directly proportional to it.

The polarimeter is a useful device for measuring the angle of rotation and is based on two Polaroid sheets arranged as in Fig 10.6. From this angle of rotation we can determine the concentration of the solution. Light from a source, usually a sodium lamp, is passed through a fixed Polaroid sheet (A in Fig 10.6), the *polariser*, through the flat end of a cylindrical cell containing the solution and through a second sheet (B in Fig 10.6), the *analyser*. On looking through the eyepiece at the analyser end two half circles of light can be seen. The analyser is rotated until both halves are of equal intensity and the reading noted. The angle of rotation for a series of known solutions is plotted against concentration and the graph used to determine unknown concentrations.

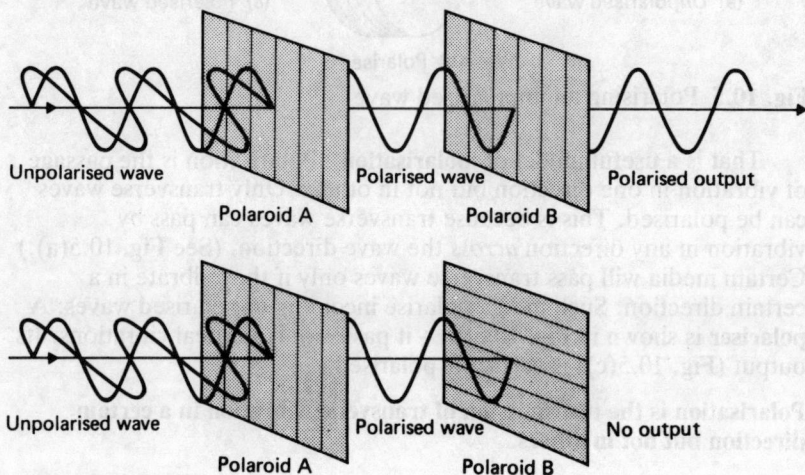

Fig. 10.6 Polaroid's effect on light

10.4 Reflection

When waves in one medium meet the surface of a second, only three things can happen.

  (i) The waves bounce back into the first medium; this is **reflection**.
 (ii) The waves pass on into the second medium; this is **refraction.**
(iii) The waves stop, the energy becoming a new form; this is **absorption.**

In most cases, a wave will be partly reflected, partly refracted, and partly absorbed. Reflection, refraction and absorption can happen with all waves.

Here are the **laws of reflection**. To keep matters simple, we discuss them for single **rays** – very narrow waves. The laws tell where a ray will travel after reflection by the surface between two media. Refer to Fig. 10.7.

1. **The reflected ray is in the same plane as the incident ray and the normal.**
2. **The angle of reflection equals the angle of incidence.**

We described the laws for a ray meeting a flat (or **plane**) surface. But they *always* apply when a wave is reflected – whatever kind of wave it is; even if the wave is wide; even if the wave is not parallel; whether the surface is smooth or rough, plane or curved.

A ray meets a reflecting surface perpendicularly. What happens to it?   [132]

A ray hits a plane mirror at 30 °. The mirror is now turned through 30 °. By how much is the reflected ray turned?   [393]

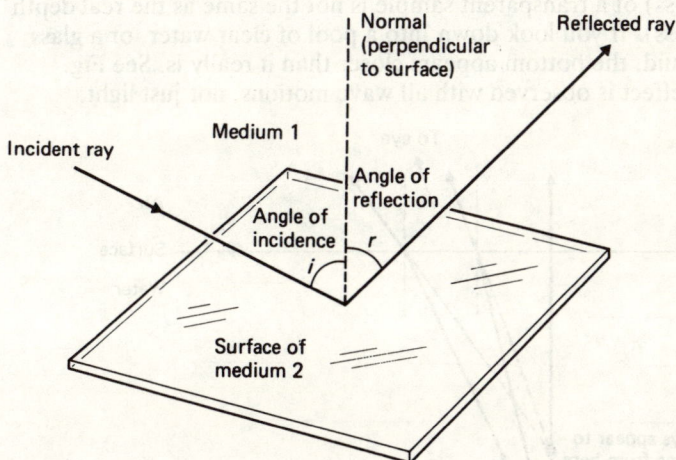

Fig. 10.7 On the laws of reflection

10.5 Refraction

A wave is refracted when it passes from one medium on into another.

What happens is that its speed changes. *In most cases its direction changes as a result.*

Refraction effects are caused by change of wave speed as the wave passes from one medium on into another.

108

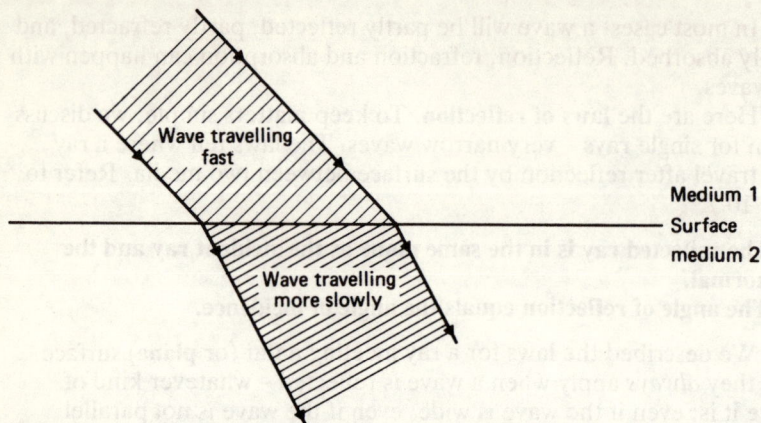

Fig. 10.8 How refraction causes a wave to bend

Figure 10.8 shows how this can cause bending; as the wavefronts change speed, they may change direction: thus sound travels more slowly in CO_2 than in air, for instance.

Here is another common effect of refraction. The apparent depth (or thickness) of a transparent sample is not the same as the real depth (or thickness). If you look down into a pool of clear water, or a glass of clear liquid, the bottom appears closer than it really is. See Fig. 10.9. The effect is observed with all wave motions, not just light.

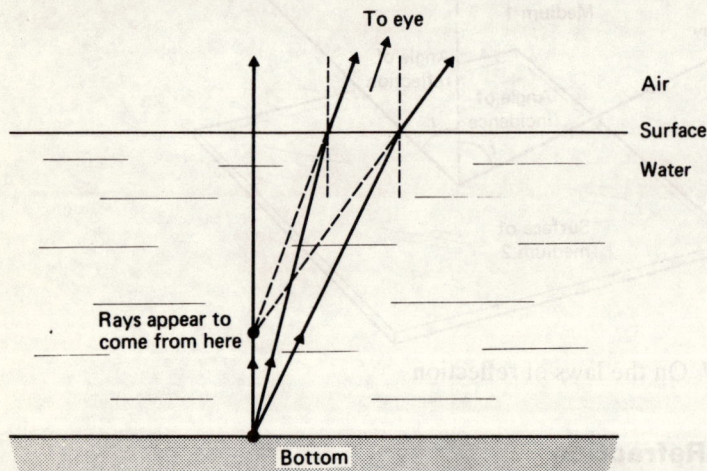

Fig. 10.9 Refraction makes a pool appear less deep than it really is

10.6 Lenses

Lenses are shaped so that the change of direction during the refraction of waves can be used.

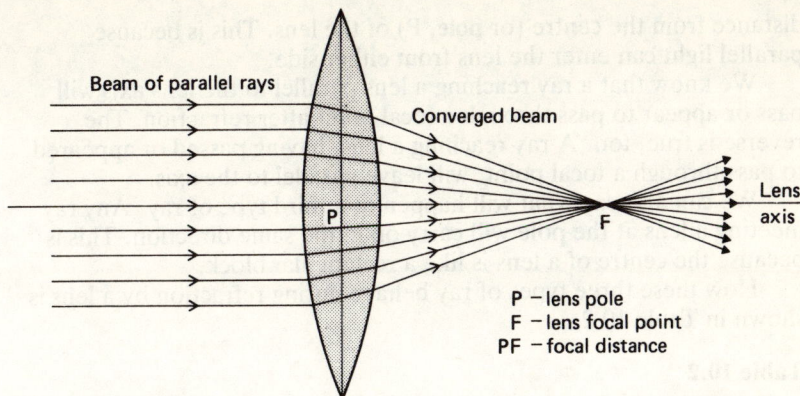

Fig. 10.10 The action of a converging lens

We can best describe lenses by their effects on an originally parallel beam of light. Some lenses concentrate, or **converge**, this. They are called **converging lenses**; normally they are thicker in the middle than at the edge. (See Fig. 10.10.)

Other lenses spread, or **diverge**, a parallel beam; **diverging lenses** are usually thinner in the middle than at the edge. (See Fig. 10.11.)

If the light leaving these lenses enters the eye, it will appear to have come from the point marked F in each case. F is a **focal point** of the lens.

The focal point of a lens is the point through which light rays, incident parallel and close to the axis, pass or appear to pass after refraction.

The distance between a lens and the focal point is called the **focal distance** (or *focal length*); its value is a measure of the **power** of a lens to converge light.

Any lens has two main focal points, one on each side, the same

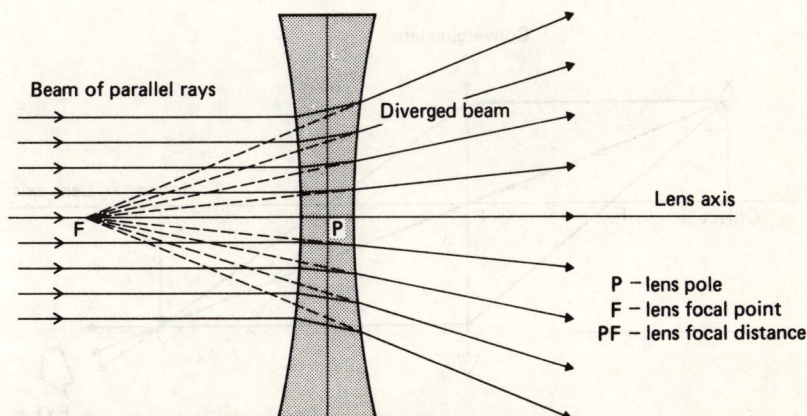

Fig. 10.11 The action of a diverging lens

distance from the centre (or **pole**, P) of the lens. This is because parallel light can enter the lens from either side.

We know that a ray reaching a lens parallel to the lens axis will pass or appear to pass through a focal point after refraction. The reverse is true, too. A ray reaching a lens, having passed or appeared to pass through a focal point, will leave parallel to the axis.

We can also say what will happen to a third type of ray. Any ray meeting a lens at the pole will carry on in the same direction. This is because the centre of a lens is like a rectangular block.

How these three types of ray behave during refraction by a lens is shown in Table 10.2.

Table 10.2

If the incident ray . . .	the refracted ray . . .
. . . arrives parallel to the axis,	. . . passes through the focal point.
. . . passes through the focal point,	. . . leaves parallel to the axis.
. . . meets the lens at the pole,	. . . carries straight on.

Figure 10.12 shows these three rays leaving the top, X, of some object, and being refracted by the lens. They leave the lens as described in Table 10.2.

If these rays enter the eye, they will appear to have come from point Y, rather than from X. Y is the **image** of X. Rays from X refracted by the lens will pass through Y; by choosing three whose behaviour is known, we can find the image of any object. (Note that all this strictly applies only to rays fairly close to the axis.)

In the case shown in Fig. 10.12, the rays entering the eye really did come from Y. Y is called a **real image**.

A real image is an image through which rays really pass.

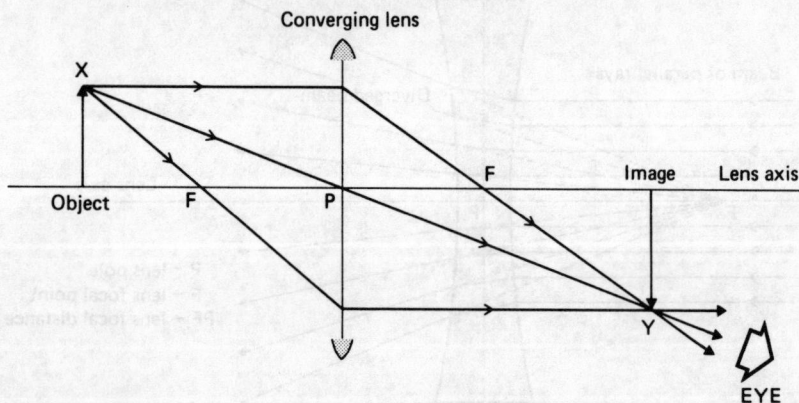

Fig. 10.12 How a lens focuses rays from an object

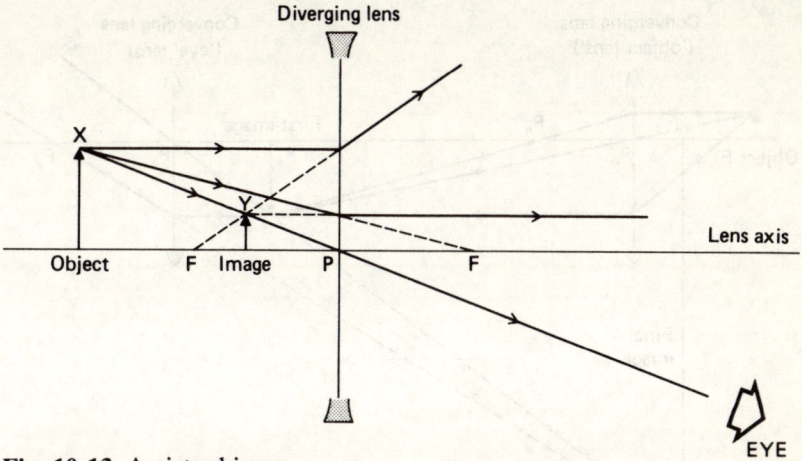

Fig. 10.13 A virtual image

Figure 10.13 shows a **virtual image** – the rays appear to come from Y, but in fact do not.

A virtual image is an image through which rays do not really pass.

10.7 Some optical instruments

Optical instruments are combinations of lenses, mirrors and prisms. They are used to produce images which can be seen better than the original objects. Telescopes and periscopes are examples.

The **magnifying glass** is an optical instrument; it is a **simple microscope**. The converging lens is used so that the small object is closer to the lens than the focal point. Figure 10.14 shows how a magnified virtual image is formed.

A **compound microscope** uses more than one lens; much more magnification can be obtained. In the simplest, two-lens, case shown in Fig. 10.15, each lens can be seen to give some magnification.

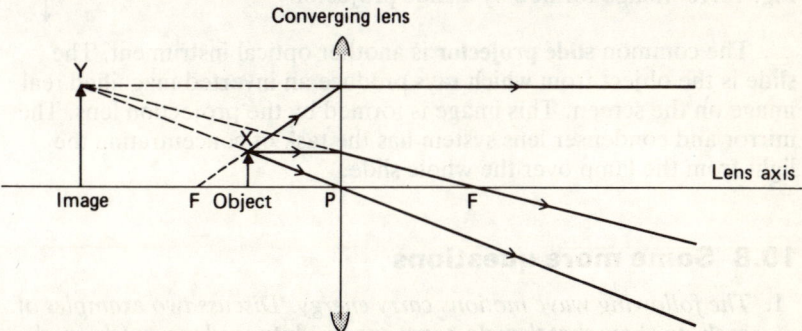

Fig. 10.14 Image formed by a simple microscope

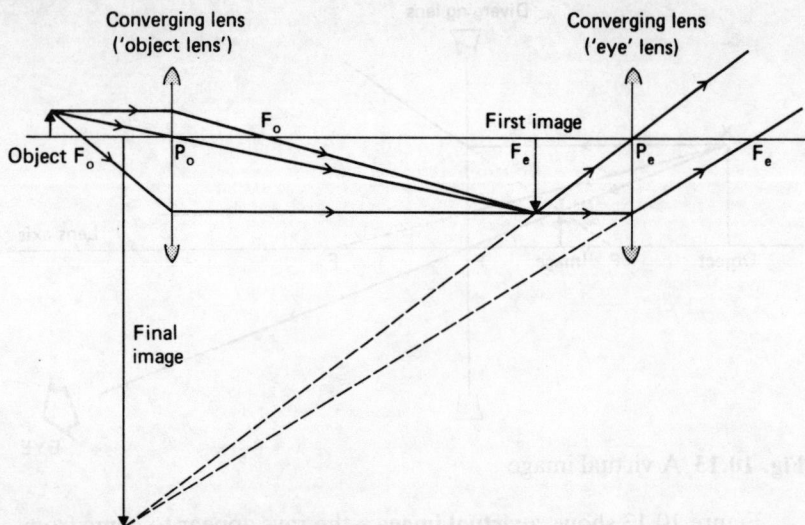

Fig. 10.15 Image formed by a compound microscope

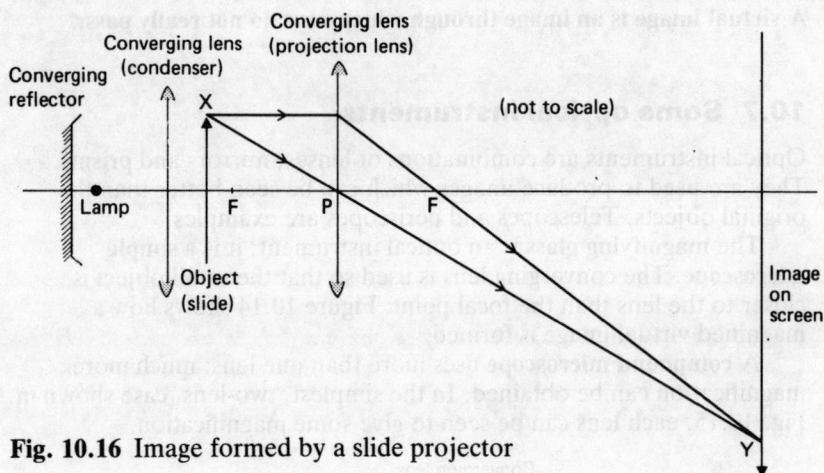

Fig. 10.16 Image formed by a slide projector

The common **slide projector** is another optical instrument. The slide is the object from which rays produce an inverted magnified real image on the screen. This image is formed by the projection lens. The mirror and condenser lens system has the task of concentrating the light from the lamp over the whole slide.

10.8 Some more questions

1. *The following wave motions carry energy. Discuss two examples of each, to show that they* do *carry energy. Infra-red waves (thermal radiation); waves on the sea from a distant storm; ultrasonic waves as used in an echosounding system; microwaves (radar waves).* [136]

2. *Medical X-rays may have frequencies around 10^{18} Hz; they travel at the speed of light. What is their period? What is their wavelength?* [226]

3. *A radar station uses waves of 3×10^{10} Hz. How many wavelengths would there be between it and an aircraft 50 km distant? What would be the time interval between a transmitted pulse and the return echo?* [381]

4. *A 1.0 km steel pipe contains water. At one end are three microphones, one each in the water, on the pipe, and in the air. Someone strikes the other end of the pipe with a hammer. After what times will the sound reach each microphone? Use Table 10.1.* [248]

5. *Explain how waves travel along a liquid surface.* [415]

6. *Explain why sound waves cannot be polarised, and how light waves can be polarised.* [338]

7. *Describe how waves in a 'slinky' spring can show polarisation (a) in the transverse case, (b) in the longitudinal case.* [153]

8. *A plane mirror stands upright, facing north, along an east-west line. A light ray travelling south-east hits the mirror. Where will the ray go after reflection? How should the mirror be turned so that the ray will go south after reflection?* [133]

9. *Explain with a diagram, why a pencil standing in a glass of water appears bent at the liquid surface.* [342]

10. *The glass lenses shown in Fig. 10.17 are used in air. Which are diverging lenses?* [328]

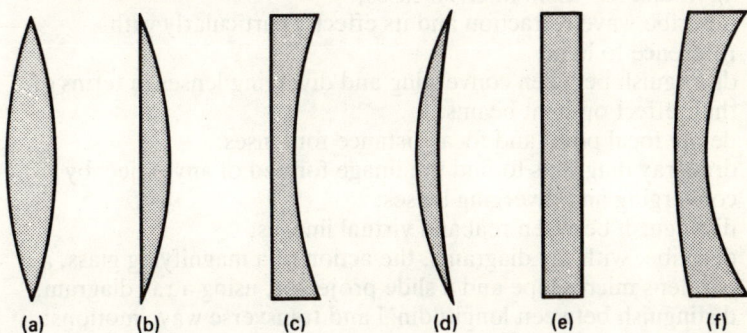

(a)   (b)   (c)   (d)   (e)   (f)

Fig. 10.17

11. *Draw accurate ray diagrams (like those in Fig. 10.12, 10.13 and 10.14) to show*
    (i) The image formed of a 10 mm-high object placed 500 mm from a converging lens of focal distance 100 mm.
    (ii) The image formed of a 20 mm-high object placed 50 mm from a diverging lens of focal distance 100 mm.
    Draw a rough ray diagram before you choose the scales to use. In each case describe the image – distance from lens, height, real or virtual, upright or inverted. [141]

12. *What happens when a magnifying lens is used to view an object beyond the focal point?* [221]

114

13. *Describe with care how you would use two pieces of Polaroid to show that light can be polarised. Explain what happens in simple terms. Why does this experiment make us believe that light travels as transverse waves?* [501]

10.9 Objectives

Note: The first nineteen objectives are from Level 1.
When you have studied this chapter, you should be able to

(1) discuss what is meant by waves as a means of energy transfer;
(2) describe sound as a pressure wave;
(3) discuss other waves;
(4) explain, with a simple sketch, the meanings of wavelength and frequency;
(5) state the frequency unit to be the hertz;
(6) state and use the basic wave equation;
(7) state that sound has a finite speed, whose value depends on the medium;
(8) state that sound is produced by vibration, and that the frequency depends on the form of the vibrator;
(9) describe experiments on the reflection and refraction of waves;
(10) state that sound waves can be reflected and refracted;
(11) state the laws of wave reflection, particularly with reference to light, and use them in given cases;
(12) describe wave refraction and its effects, particularly with reference to light;
(13) distinguish between converging and diverging lenses in terms of their effect on light beams;
(14) define focal point and focal distance for lenses;
(15) draw ray diagrams to find the image formed of any object by converging and diverging lenses;
(16) distinguish between real and virtual images;
(17) describe, with ray diagrams, the action of a magnifying glass, a two-lens microscope and a slide projector, using a ray diagram;
(18) distinguish between longitudinal and transverse wave motions;
(19) describe how these can pass along a stretched spring;
(20) give examples of each type of wave;
(21) discuss the passage of liquid surface waves;
(22) show that transverse waves can vibrate in many directions, while only one is possible with longitudinal waves;
(23) define polarisation;
(24) state why longitudinal waves cannot be polarised, while transverse waves can;
(25) describe polarisation of waves in a stretched spring;
(26) describe polarisation of light waves.
(27) deduce from this that light is a transverse wave motion;
(28) describe the use of a polarimeter to measure concentration.

Chapter 11

Interference and diffraction of waves

11.1 Overlapping waves

What happens when two waves pass through the same region of space? Do they have any effect on each other? In some sense they do not – each wave will leave the region of overlap just as if the other had not been there.

Yet in another sense overlapping waves *can* affect each other; the result, called **interference**, can sometimes be rather strange.

In the case of waves on the surface of water, the results are familiar. We are all used to the sight of small waves on the surface of larger ones. The waves add up to a single more complex wave. This is because each point on the liquid surface cannot be at more than one level at one time. Figure 11.1 shows this.

It can be seen that at any point along the picture, the actual displacement of the surface is the sum of the displacements caused by the original waves. Thus crests sum to large crests; troughs sum to large troughs; the surface is less disturbed when a crest meets a trough.

The same idea of adding displacements applies to all the waves we met in Chapter 10. In each case, the actual displacement at any point is the sum of the individual displacements (taking direction into account).

Figure 11.2 shows how two waves can be added to find the resultant wave.

The 'principle of superposition' is used here.

When two or more waves of the same type pass through the same space, the actual displacement at any point is the sum of the displacements caused by each wave.

116

(not to scale)

Swell

Waves

Ripples

Swell + waves
+ ripples

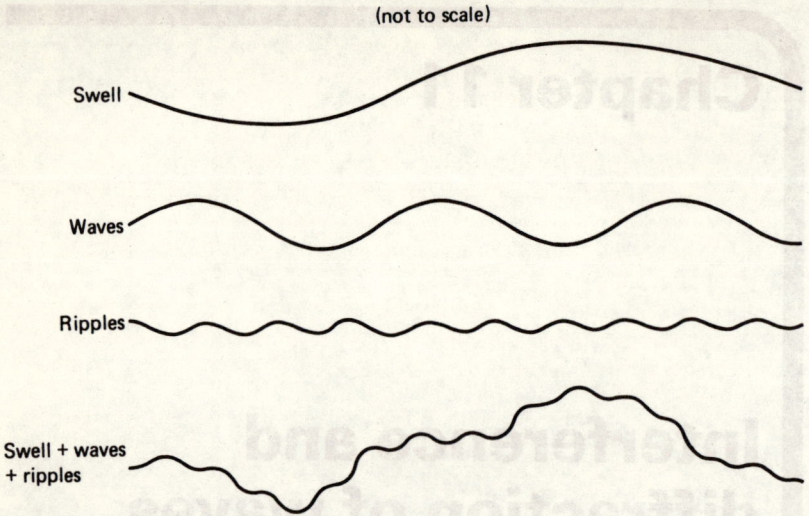

Fig. 11.1 Swell + waves + ripples (not to scale)

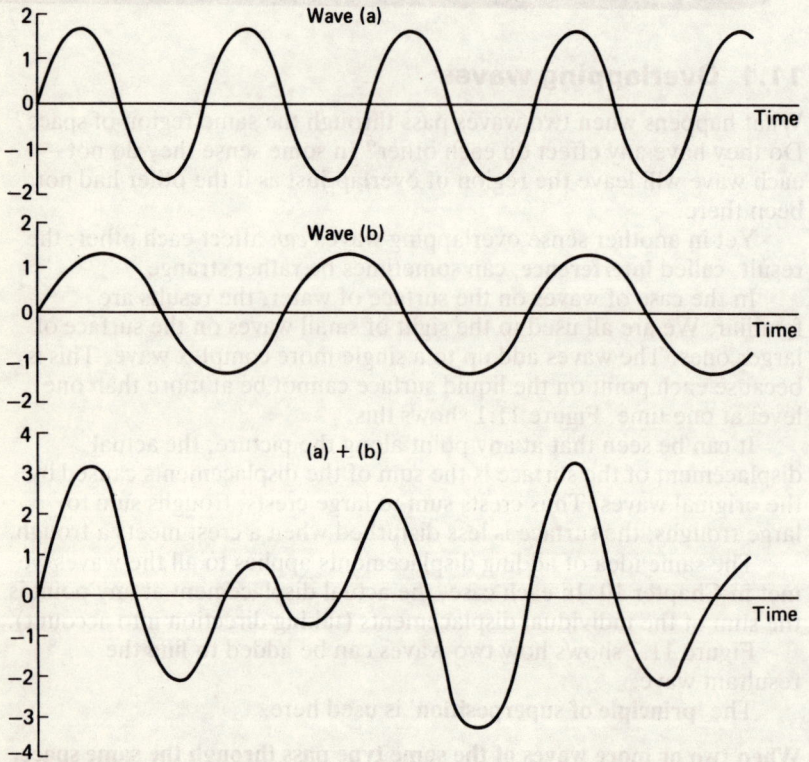

Wave (a)

Wave (b)

(a) + (b)

Fig. 11.2 Adding waves

We can use a ripple tank to investigate the striking behaviour of water waves. Such a tank is rectangular in construction with sloping sides to prevent reflection of ripples from the tank. The circular ripples are produced by the end of a vibrating steel rod. The water in the tank is at a depth of about 5 mm.

When the **ripple tank** is used to show overlapping waves, the strange effects mentioned above become clear. In Fig. 11.3 (a) we see a series of circular waves radiating from a single vibrating point. Figure 11.3 (b) shows the similar series from a second such point. Both sets of waves appear, and overlap, in Fig. 11.3 (c).

In Fig. 11.3 (c) we see points where two crests come together or where two troughs come together. XX is a row of these. As the waves pass, the liquid surface vibrates with large amplitude at these points – from double trough to double crest and back.

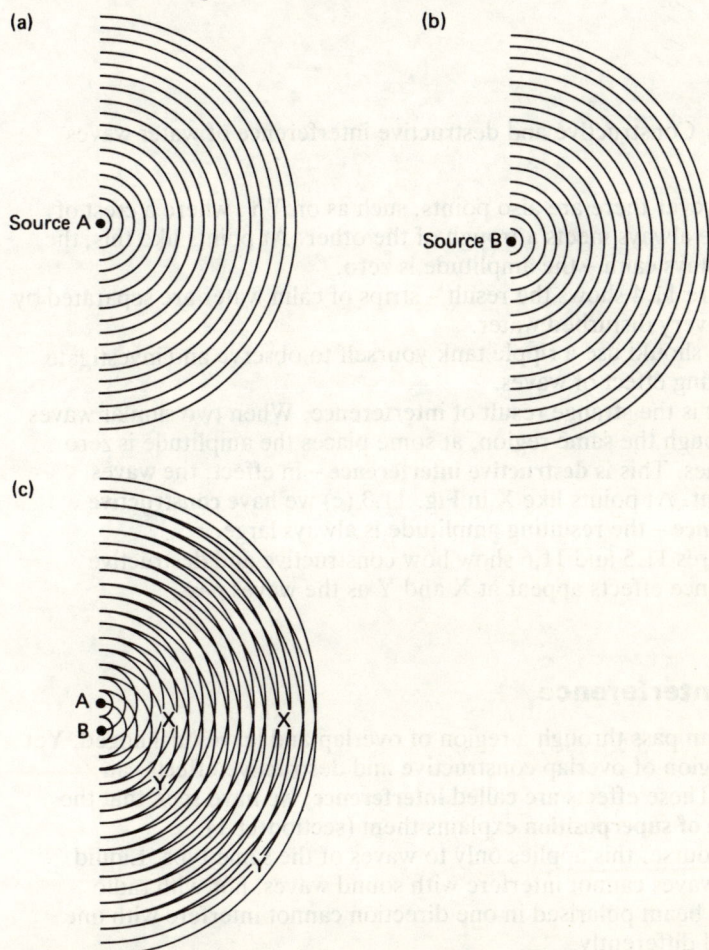

(a)

(b)

(c)

Fig. 11.3 The ripple tank shows overlapping waves

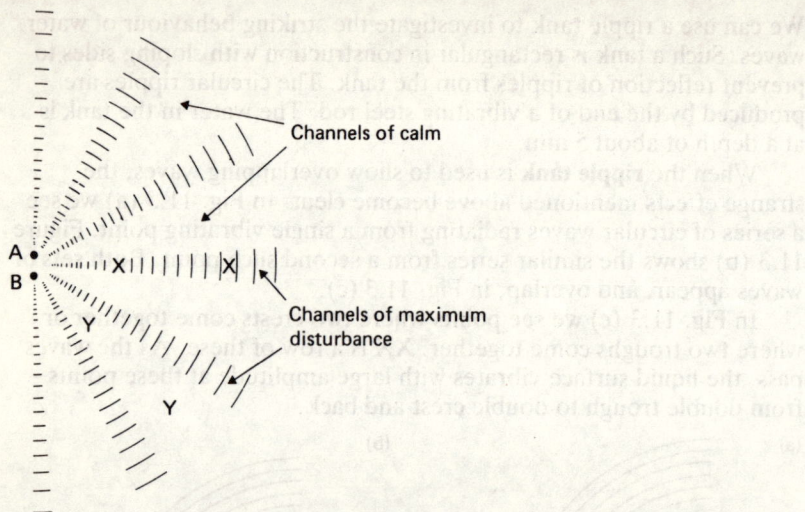

Fig. 11.4 Constructive and destructive interference of water waves

However there are also points, such as on YY, where a crest of one wave always meets a trough of the other. At points like this, the surface stays calm – the amplitude is zero.

Figure 11.4 shows the result – strips of calm water are separated by strips of very disturbed water.

You should use a ripple tank yourself to observe and investigate this striking effect of waves.

That is the strange result of interference. When two similar waves pass through the same region, at some places the amplitude is zero at all times. This is **destructive interference** – in effect, the waves cancel out. At points like X in Fig. 11.3 (c) we have **constructive interference** – the resulting amplitude is always large.

Figures 11.5 and 11.6 show how constructive and destructive interference effects appear at X and Y as the waves pass.

11.2 Interference

Waves can pass through a region of overlap and leave unchanged. Yet in the region of overlap constructive and destructive effects can appear. These effects are called **interference**; we have seen that the **principle of superposition** explains them (section 11.1).

Of course, this applies only to waves of the same type. Liquid surface waves cannot interfere with sound waves, nor with radio waves; a beam polarised in one direction cannot interfere with one polarised differently.

For clear and lasting interference patterns to appear, we must also

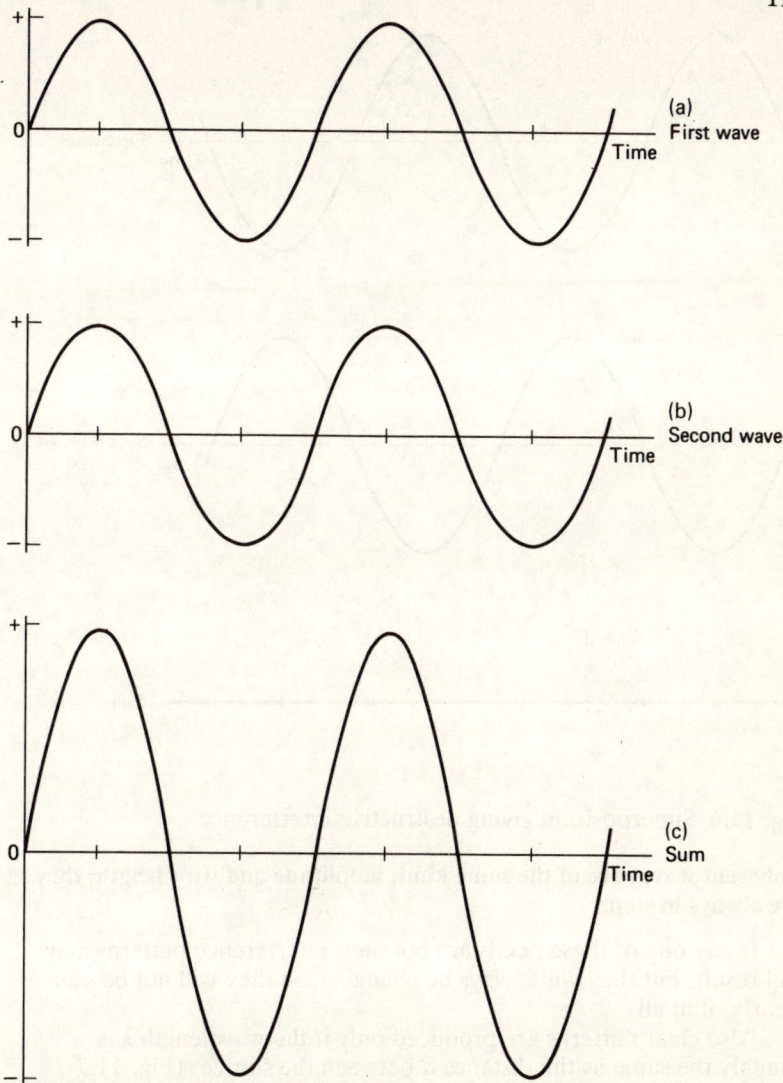

Fig. 11.5 Superposition giving constructive interference

have the following characteristics; otherwise the pattern will be confused and not easy to discuss.

1. *The waves must have similar amplitudes.* If not, there will not be full destructive interference.
2. *The waves must have the same wavelength.* If not, the interference pattern will not be static or clear.
3. *The waves must have a constant phase relationship* (they must stay in step). If not, again the pattern will not be static or clear.
   Waves like this are **coherent**.

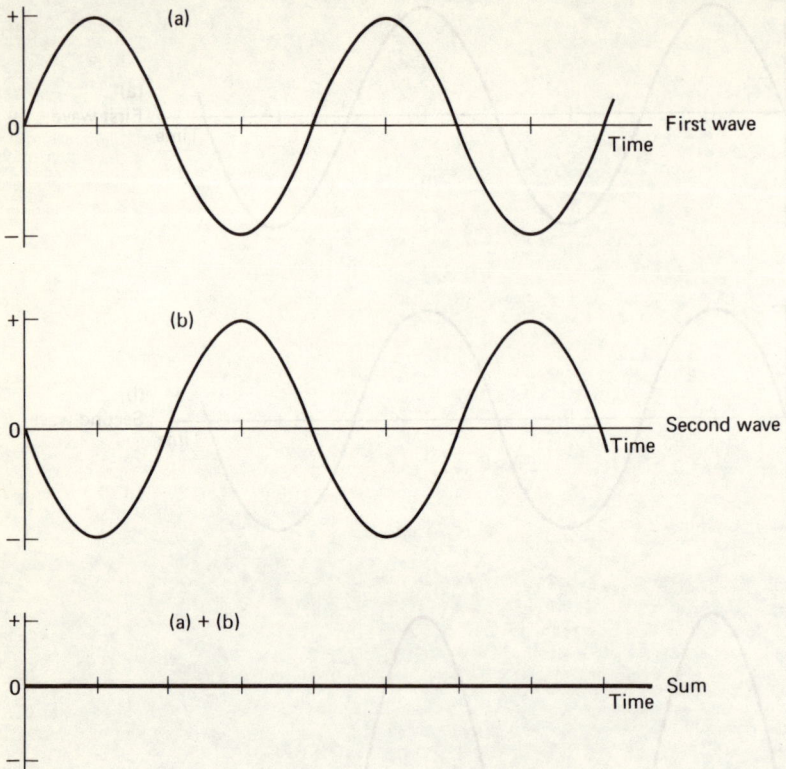

Fig. 11.6 Superposition giving destructive interference

Coherent waves are of the same kind, amplitude and wavelength; they are always in step.

If any one of these needs are not met, interference patterns may still result, but they will always be changing, so they will not be seen clearly, if at all.

Also clear patterns are produced only if the wavelength λ is roughly the same as the distance d between the sources (Fig. 11.7).

Lastly note that interference effects are easy to explain on the basis of waves. Scientists cannot explain how beams of particles could interfere in this way. Thus. **if a radiation shows interference effects, the radiation must be a wave**.

The reverse is true, too – if a radiation travels in waves, it must be able to show interference effects.

In Chapter 10 we showed that sound radiation is a wave motion. Figure 11.8 shows how sound interference effects can be produced. If you walk along the line AB (for instance), you find patches of loud sound, with patches of quiet between. Two loud sounds can add to produce quiet!

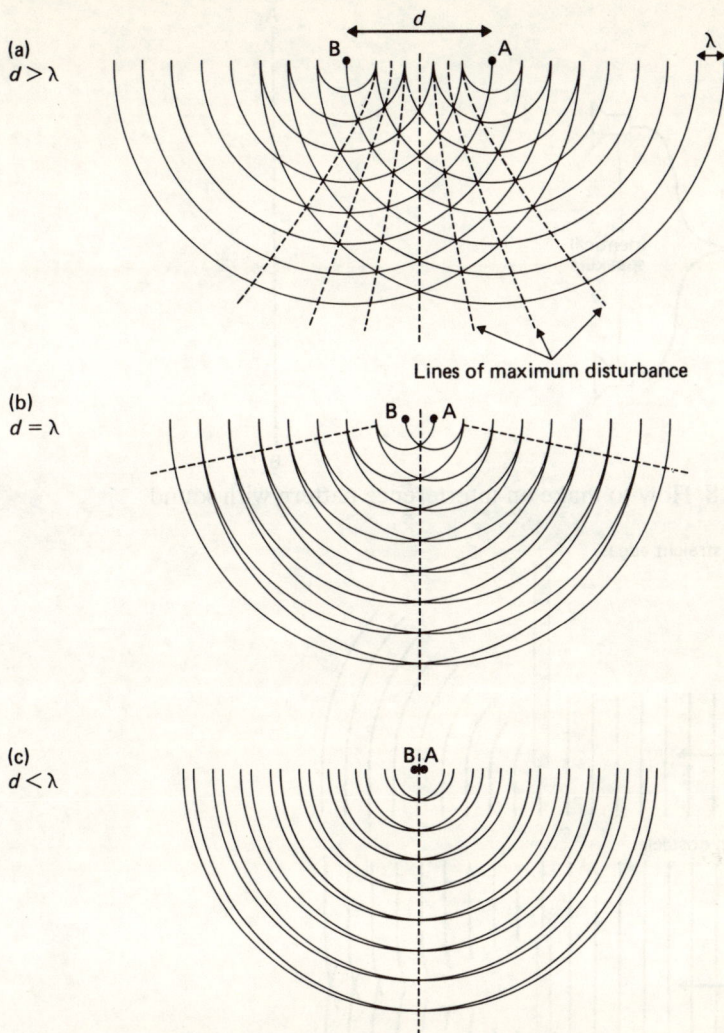

(a)
$d > \lambda$

d

B   A

λ

Lines of maximum disturbance

(b)
$d = \lambda$

B   A

(c)
$d < \lambda$

B A

Fig. 11.7 Interference patterns with various values of d/λ

11.3 Diffraction

This is a second property of certain radiations that can be explained only on the basis of waves.

Diffraction is the bending of waves around an obstacle.

Again, liquid surface waves in a ripple tank show the effects very well. Some cases are given in Fig. 11.9. Some of these can also be seen on the sea in the right conditions.

122

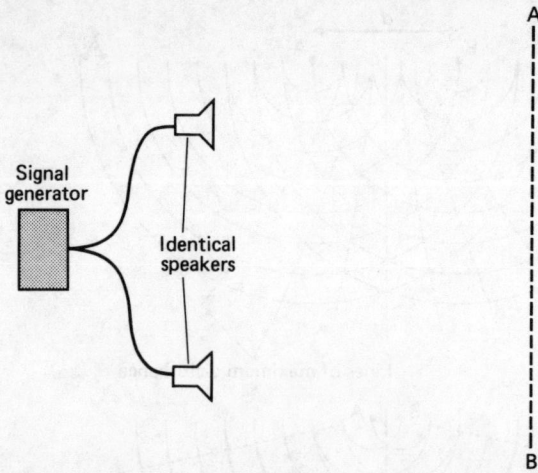

Fig. 11.8 How to make an interference pattern with sound

(a) By a straight edge:

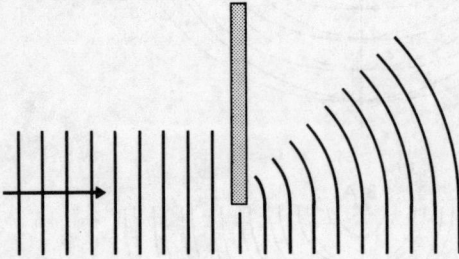

(b) By an obstacle:

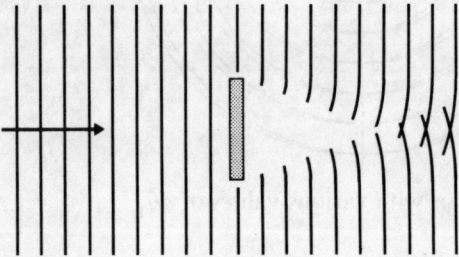

(c) By a hole:

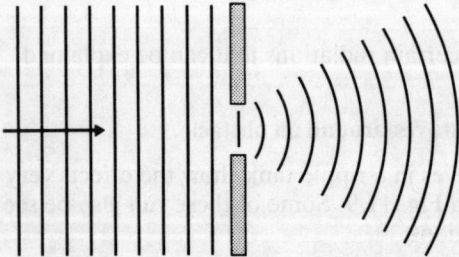

Fig. 11.9 Diffraction of water surface waves

In each case the waves bend round the obstacle into the region which we would expect to be 'in shadow' (i.e. showing no waves). Again, the effects are most clear if the obstacle (or the gap between obstacles) is about the same size as the wavelength. Again, if **a radiation shows diffraction effects, the radiation must be a wave**.

You can easily check for yourself the effects of diffraction of water surface waves using a ripple tank.

Sound waves show diffraction effects very clearly. Figure 11.10 provides an example.

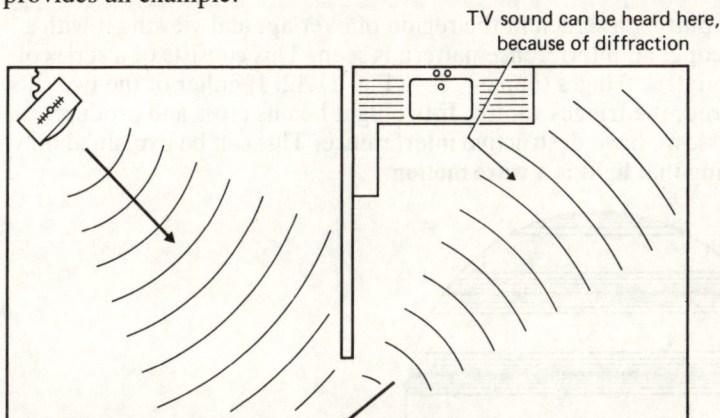

Fig. 11.10 Sound waves bend round corners by diffraction

11.4 Light is a wave motion

This was one of the major findings at the start of the nineteenth century. By showing that light could produce interference patterns, Thomas Young proved that light travels in waves.

Young's experiment is done by shining light of a single colour ('monochromatic' light) on to two very close slits. (See Fig. 11.11.)

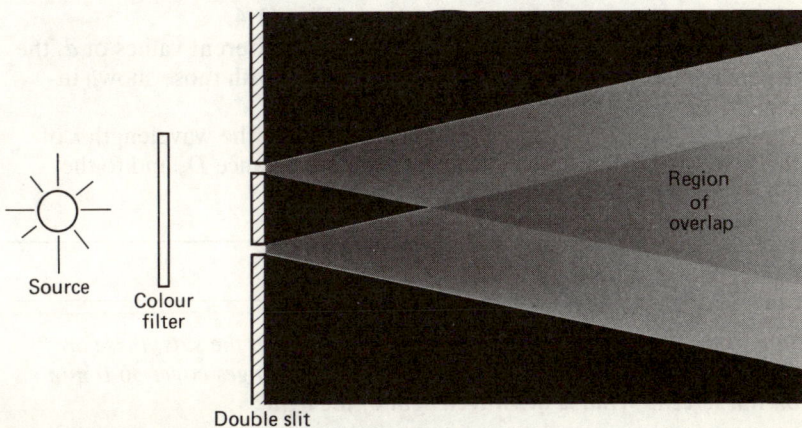

Fig. 11.11 Young's two-slit experiment

Each slit is now a light source. In the region of overlap, the two beams meet the conditions for clear and lasting patterns (compare section 11.2):

(i) The beams are the same type. (ii) The beams have the same brightness. (iii) The beams have the same colour. (iv) The beams are always in step.	The beams are coherent.

By putting a screen in the region of overlap, and viewing it with a microscope, an interference pattern is seen. This consists of a series of light and dark **fringes** (bands), as in Fig. 11.12. If either of the two slits is covered, the fringes vanish. If two light beams cross and produce darkness, we have destructive interference. This can be explained only by stating that **light is a wave motion**.

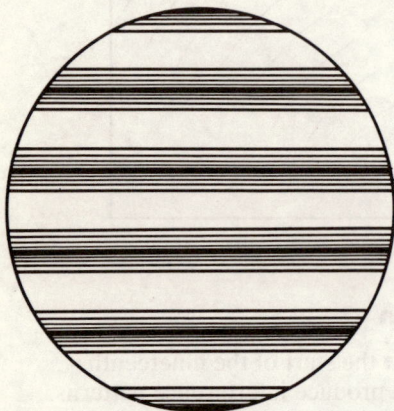

Fig. 11.12 Light interference fringes (view with half-closed eyes)

Clearly the arrangement is much the same as that of Fig. 11.3; the fringe pattern is much the same as that of Fig. 11.4.

The two-slit experiment can be tried with different values of d, the distance between the slits; the results compare with those shown in Fig. 11.7.

In fact this can be taken further, to measure the wavelength λ of the light. This relates to d, to the slit-screen distance D, and to the distance y between two bright fringes.

$$\lambda = y \times d/D$$

In a two-slit experiment, the screen is 2.00 m from the slits; these are 0.4 mm apart. In the fringe pattern, ten bright fringes cover 36.0 mm on the screen. What is the wavelength of the light?

$y = 36.0/9$ mm $= 4.0 \times 10^{-3}$ m $d = 0.40 \times 10^{-3}$ m
$D = 2.00$ m $\lambda = ?$

$\lambda = y \times d/D = 4.0 \times 10^{-3} \times 0.40 \times 10^{-3}/2.00$ m
$= 0.80$ μm $= 800$ nm

In the same experiment, the screen is moved 1.00 m closer to the slits.
What will be the effect? [148]

Work like this shows that light is a wave motion of wavelength around
500 nm (0.5×10^{-6} m).

Is light a transverse or a longitudinal wave motion? Recall the
effect of polarisation – section 10.3. This shows that light travels in
transverse waves, for only transverse waves can be polarised.

11.5 The diffraction grating

Young's two-slit experiment shows that light travels in waves. In that
case, light should also show diffraction effects (section 11.3).

That is indeed the case. The results sketched in Fig. 11.9 for water
surface waves can all be obtained with light beams.

Thus, even with a point source of light, shadows are not really
sharp at the edges – the light bends round the obstacles to some extent.
Indeed, the shadow of a small steel ball has a tiny bright spot in the
centre.

However, the diffraction of light is not as easy to discuss as that of
water waves. In most cases patterns are produced rather than simple
bending of the waves. Diffraction patterns are like interference
patterns. In fact they *are* interference patterns – dark and light fringes;
the various diffracted rays overlap and interfere.

Figure 11.13 gives an example. A beam of parallel light shines onto
a single slit (Fig. 11.13(a)). We would expect a sharp line of light on
the screen – as in (b) – if not for diffraction. But the line we see is not
fuzzy at the edges – it does not look like (c); it shows a fringe pattern
as in (d).

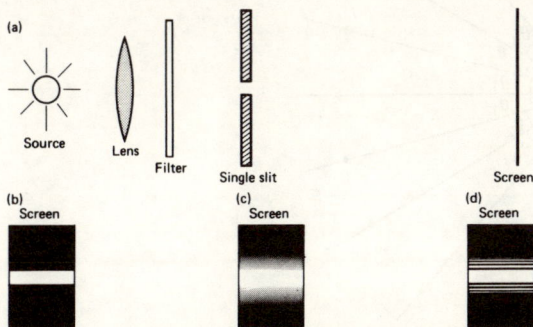

Fig. 11.13 Single-slit diffraction

Diffraction patterns are the result of interference effects.

If the light in Fig. 11.13(a) shines onto more than one slit, the pattern on the screen changes; the interference situation (caused by diffraction) becomes more complex. Figure 11.14(a) shows the pattern from five close slits; (b) is the product of ten.

We see here a growing concentration of light into a few directions; the angle of each depends on the wavelength.

The **diffraction grating** does this even more. The normal type is a thin piece of flat glass; onto one surface a large number of parallel, close grooves are cut. Light can pass only through the clear strips between the grooves. In effect it is passing through a number of parallel, close slits.

A light beam striking the grating will diffract into beams at various angles θ (though some will pass straight through). (See Fig. 11.15.)

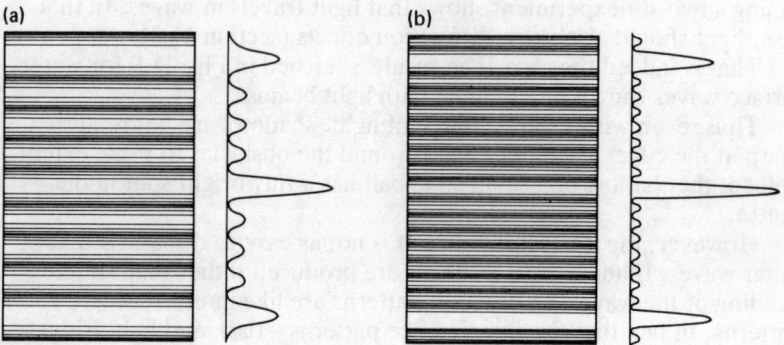

Fig. 11.14 Multiple-slit diffraction

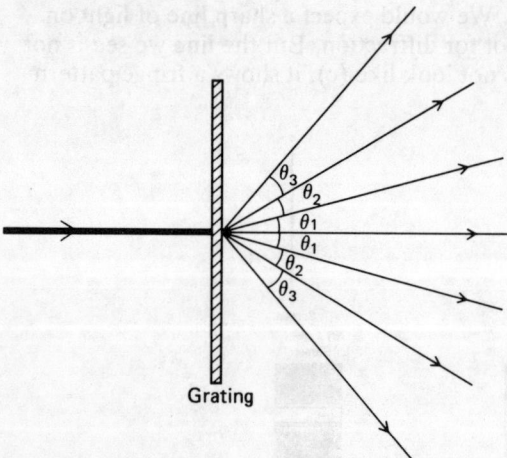

Fig. 11.15 The effect of a grating

All this has been described for light of a single colour, although we said that the output beam angles depend on wavelength.

What happens if white light meets a grating? The light of each wavelength will split into beams as in Fig. 11.15 – but the values of θ in each case will differ. The greater the wavelength, the greater each value of θ. This effect is called **dispersion**. (There is more on this in the next chapter.)

Dispersion is the bending of radiation at an angle which depends on wavelength.

So what happens if white light meets a grating? A series of colour **spectra** will be produced. Only the light that passes straight through will remain white.

What we call white light is thus a mixture of radiations of different wavelengths; we see these as different colours. (See Chapter 12.)

11.6 Some more questions

1. *Figure 11.16(a) and (b) show two waves. Copy them with care and construct the result of their passing the same point.*   [134]

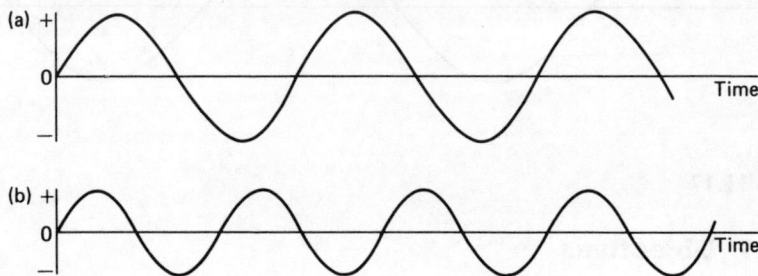

Fig. 11.16

2. *Figure 11.17(a) shows the result of two waves passing the same point. One of those waves is given in Fig. 11.17(b). Copy the two waves with care and construct the other wave.*   [361]
3. *Copy Fig. 11.8 to a suitable scale. The speakers are 0.5 m apart; the wavelength is 0.5 m; AB is 2.0 m from the speakers. Add to your drawing the waves from each speaker. What is the distance between two loud patches? Check your answer using the Young two-slit relation.*   [139]
4. *In a Young's two-slit experiment, the slits are 1.0 mm apart and 1000 mm from the screen. What will be the fringe spacing with light of wavelength 500 nm?*   [172]
5. *What would be seen on the screen in a Young's interference experiment if the two slits are bathed in white light?*   [193]

128

6. *The type of grating described in this chapter is a transmission grating – the light passes through it. Reflection gratings are sometimes used. Using sketches, explain how you think these act.* [262]

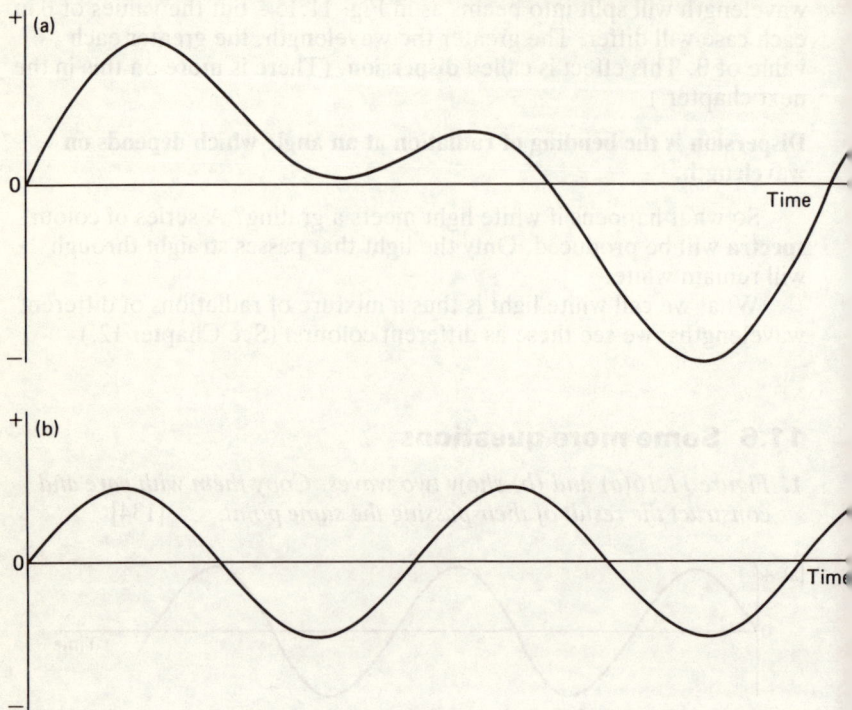

Fig. 11.17

11.7 Objectives

When you have studied this chapter you should be able to
(1) explain the nature and cause of constructive and destructive interference with liquid surface waves;
(2) sketch interference patterns obtained with liquid surface waves;
(3) state the principle of superposition and use it with simple sine waves;
(4) describe liquid surface wave diffraction patterns of various kinds;
(5) describe Young's two-slit experiment, its results, and its significance;
(6) use the wavelength relation applicable to this experiment when carried out with monochromatic light;
(7) describe coherent waves;
(8) describe single-slit and multiple-slit diffraction patterns;
(9) describe and simply explain the dispersion of light from different sources by a grating;
(10) state that white light is a mixture of radiations of different wavelengths.

Chapter 12

Spectra

12.1 The dispersion of light

Dispersion was defined at the end of the last chapter.

Dispersion is the bending of radiation at an angle which depends on wavelength.

We saw then that white light can be dispersed using a **diffraction grating**. The values of θ in Fig. 11.15 depend on the wavelength λ of the light. As **white light consists of many wavelengths**, a grating splits it into a 'spectrum'. Spectra are the subject of this whole chapter. They can be produced from all types of radiation – even particles. In each case the beam of radiation is split up into different parts (often on the basis of energy) and each part appears in a different place.

In the case of visible light, the parts with different energy have different wavelengths; we see these as different colours. So the spectrum of visible light is a band of different colours, perhaps on a screen.

We saw, too, in Fig. 11.16 that the values of θ are greater for longer waves. The end of each grating spectrum nearer the axis consists of short-wavelength light; the distant ends are long-wavelength light.

When we do the experiment, white light spectra appear as in Fig. 12.1. You should carry out this experiment yourself, perhaps with a spectroscope.

Thus 'red' light has long wavelengths; 'blue' light has short wavelengths.

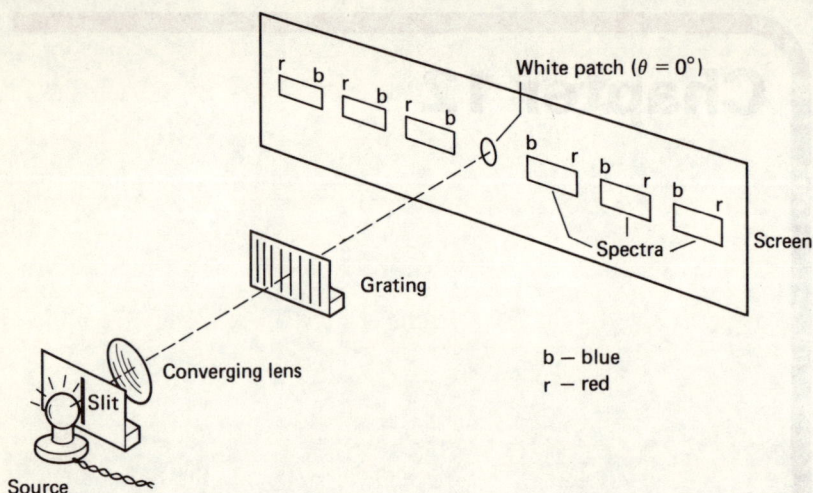

Fig. 12.1 White light spectra with a grating

Recall (a) the wave equation, and the meaning of frequency, ν (section 10.1); (b) that the speed of light in empty space (or in air) is 3×10^8 m s^{-1}.

Some red light has wavelength 750 nm. What is its frequency?

$\lambda = 750 \times 10^{-9}$ m $\quad c = 3.00 \times 10^8$ m s^{-1} $\quad \nu = ?$
$c = \nu\lambda \rightarrow \nu = c/\lambda = 3.00 \times 10^8/750 \times 10^{-9}$ Hz $= 4 \times 10^{14}$ Hz

Some blue light has wavelength 380 nm. Find its frequency. What is the frequency of light of twice the wavelength? [179]

A second way to disperse light is by use of a glass **prism**. (This is a three-sided block.) Now the dispersion is the result of refraction (see section 10.5). The rays travel through the glass at speeds which depend on wavelength; they then leave the prism at different angles. Figure 12.2 shows this; refer also to sections 10.5 and 10.6.

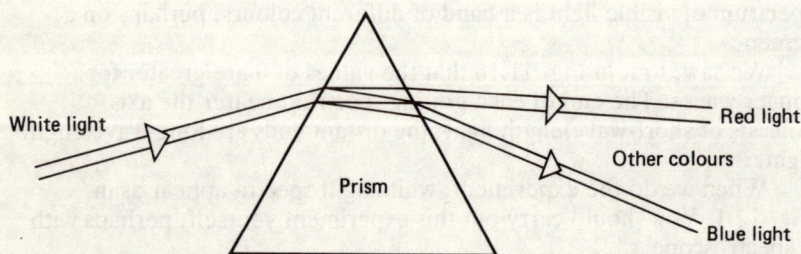

Fig. 12.2 Dispersion by a prism

The white light, moving at 3×10^8 m s^{-1}, passes into the glass. Here its speed is less and the longer wavelengths are slowed more than the shorter ones. Thus the blue part of the light is bent more than the red; the other colours lie in-between. The same happens as the waves leave the glass – the blue speeds up more than the red, so it bends more.

Figure 12.3 outlines how to set up a prism to produce a good spectrum. Light from the whitehot source is made into a narrow parallel beam by the slit and converging lens. Light of each wavelength leaves the prism as a narrow parallel beam; the angle through which it has bent depends on the wavelength. The second converging lens brings each parallel beam to focus on the screen. Light of each wavelength will come to a different point.

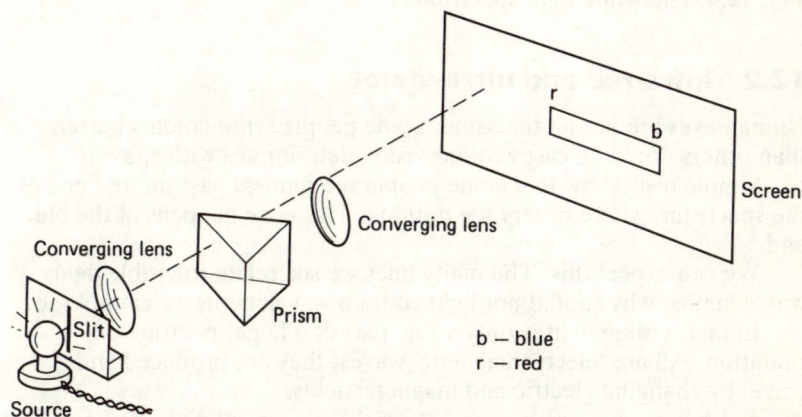

Fig. 12.3 White light spectrum with a prism

This time we obtain a single spectrum from the white light beam. Red wavelengths have bent least; blue wavelengths have bent most.

What does a white light spectrum look like? It is **a continuous band of colour**, shading from a very deep red, through yellows and greens, to a very deep blue (violet). The normal human eye can detect a couple of hundred colours ('**hues**') in a good spectrum. However, for convenience different regions are given different colour names. Table 12.1 gives the details.

Table 12.1 Regions of the white light spectrum

Wavelength/nm	400–420	420–450	450–500	500–600
Name	violet	indigo	blue	green

	560–600	600–650	650–760
	yellow	orange	red

It is not easy to draw a white light spectrum. Figure 12.4 is an attempt; on it are given the convenience names of the sections in Table 12.1. Do *not* forget that really the white light spectrum is a band of hundreds of hues; these shade gently into each other. Even each hue is really a range of wavelengths of a few nanometres.

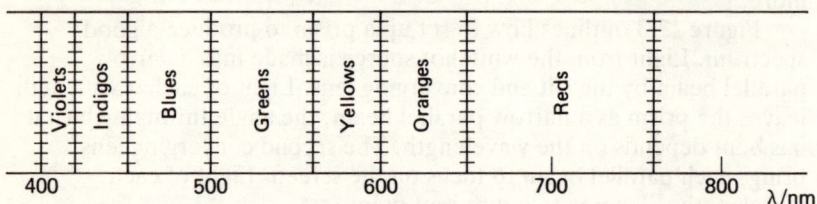

Fig. 12.4 The white light spectrum

12.2 Infra-red and ultra-violet

Human eyes are not all the same. Some people sense colours better than others do; in all cases colour vision deteriorates with age.

Simple tests show that some people see redness past the red end of the spectrum, where others see nothing. The same happens at the blue end.

We can expect this. The many hues we see relate to visible light wavelengths; why should not light contain wavelengths we cannot see?

In fact, visible light is only a tiny part of a huge spectrum of radiation. All are 'electromagnetic' waves; they are produced and travel by changing electric and magnetic fields.

Visible light is a type of electromagnetic radiation; the visible light spectrum is part of the electromagnetic spectrum.

We shall return to this in section 12.4. Now we discuss the neighbours of visible light in the electromagnetic spectrum – the radiations past the red and blue ends in Fig. 12.4.

Infra-red radiation is the long wavelength neighbour of visible light, past the red end of the visible spectrum.

Infra-red is not novel. If you switch an electric ring on, you can detect it before the ring starts to glow red. **Infra-red is 'thermal' radiation**. Our 'heat'-sensitive skin cells can detect it; so can a thermometer (section 9.1). So infra-red carries energy and can be absorbed (compare with section 10.1).

Infra-red shows all the wave properties discussed in the last two chapters – reflection, refraction, diffraction, interference. Infra-red waves can be polarised, so they are transverse. They travel at 3.00×10^8 m s^{-1} in empty space and in air.

In fact, infra-red is just like visible light, except for the longer wavelengths. To work with it requires other techniques – quartz rather than glass for lenses and prisms, thermometer rather than eyes for detection – but the results are the same.

An infra-red wavelength is 3 nm. What is its frequency? [271]

In the same kind of way ultra-violet radiation is the short wavelength neighbour of visible light, past the blue end of the visible spectrum.

These waves cannot be sensed by any special cells in humans; however they can cause the skin colour to darken and can damage skin and eyes.

A useful property is **fluorescence** – some substances glow visibly when ultra-violet light is absorbed. Discharge lamps ('fluorescent' lamps) produce ultra-violet rays; the glass is coated with fluorescent powder to convert this to light we can see. Many paints and detergents contain such materials to increase the visible brightness.

Ultra-violet carries energy in the form of transverse waves. It is just like visible light, except for the shorter wavelengths.

A 10^{16} Hz ultra-violet wave travels through a substance at 1.5×10^8 m s^{-1}. Find its wavelength. [316]

So now we can extend Fig. 12.4.

Note the unusual scale of the spectrum drawn in Fig. 12.5 – the steps are *powers* of ten rather than multiples. This is a 'logarithmic' system, used where a very large range must be plotted.

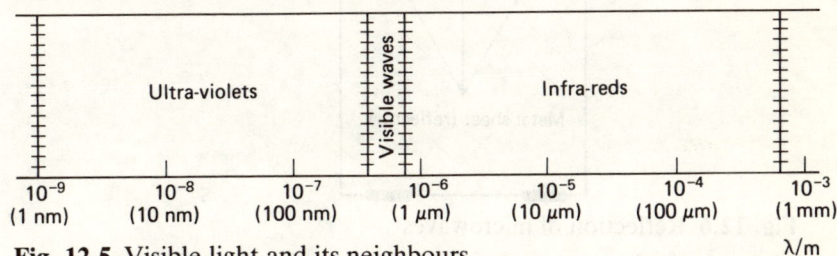

Fig. 12.5 Visible light and its neighbours

12.3 Radio waves

Beyond the infra-red section of the electromagnetic spectrum is the radio spectrum. The section of this closest to the infra-red is the **microwave** region. Microwaves are used in radar, so some people call them radar waves. Special fast cookers work with microwaves instead of infra-red.

You may have seen experiments on microwaves. The ones noted here are designed to show that these radiations are very similar to light. They need a microwave transmitter and a microwave detector (receiver). The needle on the detector meter measures the strength of the microwaves. Straight away this shows that **microwaves carry energy and can be absorbed.**

Microwaves can be reflected. (See Fig. 12.6.)

Microwaves can be refracted.

Microwaves show diffraction effects. Attempt to produce the effects of diffraction of water surface waves shown in Fig. 11.8, using microwaves. You will need to use a microwave transmitter and detector like those in Fig. 12.6 and a selection of metal objects. You will find that these must have similar dimensions to the wavelength of the radiation in order to get positive results (see section 11.3).

Microwaves show interference effects. Here model the two slit case of Fig. 11.11 using a probe detector of microwaves rather than the large one of Fig. 12.6. Use the Young two-slit relation to obtain a value for the wavelength of the waves used (see section 11.4).

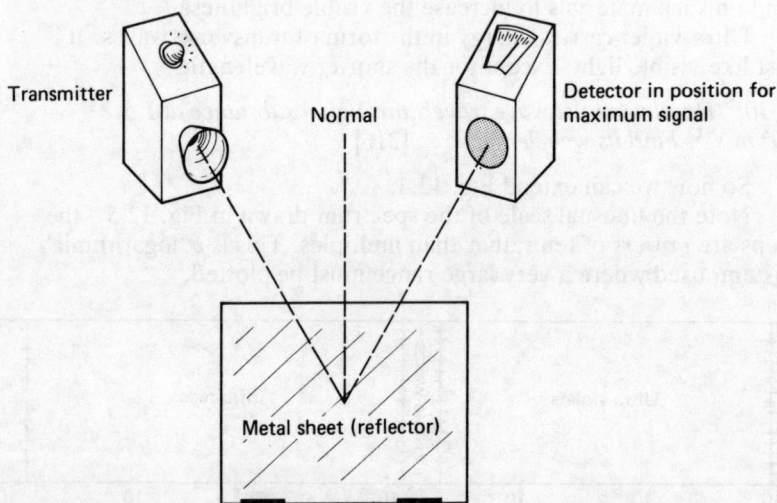

Fig. 12.6 Reflection of microwaves

Also, **microwaves can be polarised**. In fact, the transmitter shown produces polarised microwaves, and the detector detects only vibrations in one plane. So if the detector is put right by the transmitter but on its side, the meter will read zero.

It is not easy to measure the speed of radio or microwaves. However, the value found is 3.00×10^8 m s^{-1}.

Radio travels in transverse waves, with just the same properties as light. Radio radiations are part of the electromagnetic spectrum.

In a certain microwave experiment, these values were found. The distance between two slits in front of the source was 100 mm; the slit-detector distance was 250 mm; constructive interference was found at points 75 mm apart. Find the wavelength and frequency of the waves. (See section 11.4.)   [362]

12.4 The electromagnetic spectrum

People have tried tests like those in the last section with many radiations. The electromagnetic spectrum includes all those with the same properties as light.

Here again are those properties.

Radiation – carrying energy, absorbed by matter to some extent.
Wave form – showing reflection, refraction, interference, diffraction.
Transverse vibration – can be polarised.
Speed – 3.00×10^8 m s^{-1} in empty space and in air.

Electromagnetic radiations are transverse waves that travel at 3.00×10^8 m s^{-1} in empty space and in air.

We have also noted that they are produced and carried by changing electric and magnetic fields. (See Fig. 12.7.) This is the reason for their name – electromagnetic.

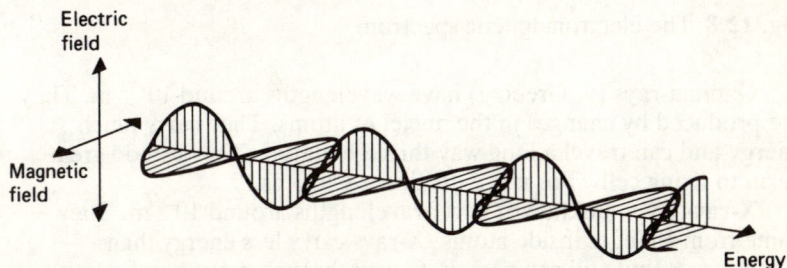

Fig. 12.7 An electromagnetic wave

The vibrating electric and magnetic fields of these radiations result from moving electric charges (Chapter 14). The two fields interact with each other and carry energy along at 3×10^8 m s^{-1} in empty space.

The electromagnetic waves we have met so far in this chapter are listed in Table 12.2. The wavelength limits are chosen for convenience.

Copy the table with a third column – frequency/Hz. Work out and enter the values that relate to the wavelengths quoted.   [371]

Table 12.2 Some electromagnetic radiations

Radiation name	Wavelength range/m
Ultra-violet	10^{-9} to 4×10^{-7}
Visible light	4×10^{-7} to 7.6×10^{-7}
Infra-red	7.6×10^{-7} to 10^{-3}
Microwave	10^{-3} to 0.12
Radio	0.12 to 10^4 or more

Now we can show the whole electromagnetic spectrum – Fig. 12.8. Note again two points. (a) The wavelength values that divide the spectrum into bands are for convenience only – really the radiations merge into each other. (b) A logarithmic scale is used as in Fig. 12.5 as the spectrum is so long.

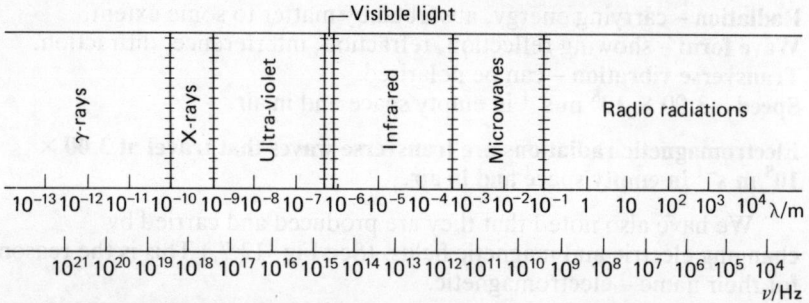

Fig. 12.8 The electromagnetic spectrum

Gamma-rays (γ, Greek g) have wavelengths around 10^{-12} m. They are produced by changes in the nuclei of atoms. They carry much energy and can travel a long way through matter. They can do great harm to living cells. For more detail see Chapter 19.

X-rays (X = unknown) have wavelengths around 10^{-8} m. They come from changes inside atoms. X-rays carry less energy than gamma-rays, but still can pass quite well through matter and cause harm to living cells.

Ultra-violet and **visible light** also come from changes inside atoms. They do not pass well through matter and do little harm.

Infra-red (thermal) radiations are made by changes in large particles and groups of particles.

Microwaves (radar waves) can arise in the same way as infra-red, or from changing electric currents in special devices.

Radio waves result from very large changes, such as sparks and changing currents in aerials. Their main use is in telecommunications; however, all electromagnetic radiations except gamma and X-rays have this use now.

12.5 Line spectra

We have seen that different electromagnetic radiations result from electromagnetic changes on different scales in matter.

To take this further some simple ideas of the structure of matter are needed. They are given in the box; we deal with them in more detail in the last chapter of this book. (See also section 6.1.)

(i) All matter consists of **particles**.
(ii) The basic particle is the **atom**.
(iii) Each atom has a small central **nucleus**; this carries a positive electric charge. A cloud of negative charge surrounds the nucleus, carried by negative **electrons**.
(iv) Atoms tend to join with each other; one result of this is **molecules**.
(v) In this book the phrase 'particles of matter' usually relates to atoms or molecules. In **metals**, however, it includes the many 'free' electrons.

These ideas have been kept simple – they are less complex than in reality.

Matter emits electromagnetic radiation whenever electric charge accelerates in it. The wavelength of the radiation relates to the size of the source.

A major advance in physics this century was the *quantum theory*. This accounts for very many facts; they include how electromagnetic radiations are produced and how they behave.

The quantum theory states that the energy of a system is restricted to certain values.

We call these values 'allowed quantum states'. The system is not allowed to have other values – the other values are 'forbidden quantum states'. The idea is one of the most useful in modern physics.

For instance, the nucleus of an atom is allowed only certain energies. In each case, we can show these on an energy level diagram like that in Fig. 12.9. The actual energy of the nucleus at any time is shown on the diagram by a 'blob' at the correct level. The lowest allowed energy state for the system is the **ground state**, W_o.

Thus in Fig. 12.9(a) the nucleus is in the ground state. In (b) its energy is above W_o – the nucleus is **excited**.

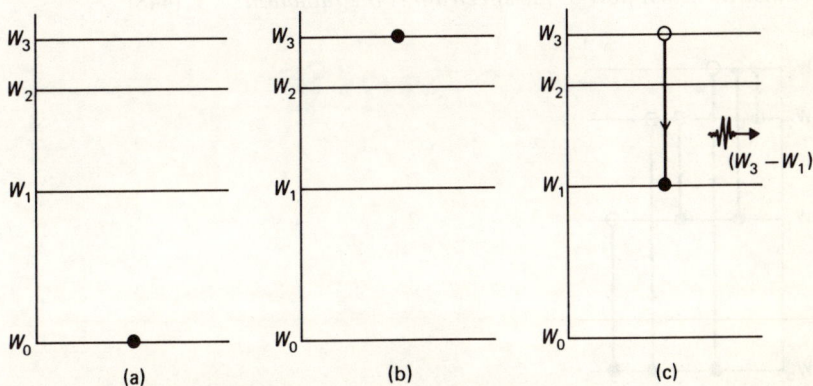

Fig. 12.9 Quantum theory energy level diagrams

138

An excited system will tend to release the excess energy and 'fall down' towards the ground state. When such a change occurs, the energy released appears as a **quantum** of electromagnetic radiation. (Quanta are often called **photons** in this case.)

A quantum is a single 'packet' of energy.

Thus in Fig. 12.9(c), the system falls from level W_3 to level W_1. A quantum of radiation appears; its energy is $(W_3 - W_1)$. Here it will be a gamma-quantum, or gamma-photon.

The quantum theory also relates the energy of a quantum to its frequency.

Quantum energy = Planck constant × frequency
$W = h\nu$.

The **Planck constant**, h, is very important in the quantum theory. It is the ratio of the energy of a quantum to its frequency:

$h = W/\nu$ Unit: joule second, J s.

Its value is found in the next question.

A gamma-photon of frequency 10^{22} Hz carries 6.6 pJ. What is the value of the Planck constant?

$\nu = 10^{22}$ Hz $W = 6.6 \times 10^{-12}$ J $h = ?$

$W = h\nu \rightarrow h = W/\nu = 6.6 \times 10^{-12}/10^{22}$ J s $= 6.6 \times 10^{-34}$ J s.

What energy is carried by a 500 nm light photon?   [401]

An atom falls between levels of energy 2.35×10^{-17} J and 3.35×10^{-16} J. Find the energy, frequency and wavelength of the quantum it emits. In which part of the spectrum is the radiation?   [448]

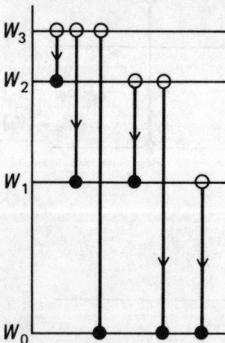

Fig. 12.10 Possible decays in a system with three excited states

Now look again at Fig. 12.9, the sample energy level scheme. If the system concerned is excited to level W_3, (b), the fall to W_1, (c), is not the only one that could happen. Figure 12.10 shows *all* the ways in which this system could decay after excitation.

So a nuclear system with three excited states can emit these quanta:

Table 12.3 Quanta produced by a system with three excited states

Initial level	Final level	Released energy	Quantum frequency
W_3	W_2	$W_3 - W_2$	$(W_3 - W_2)/h$
W_3	W_1	$W_3 - W_1$	$(W_3 - W_1)/h$
W_3	W_0	$W_3 - W_0$	$(W_3 - W_0)/h$
W_2	W_1	$W_2 - W_1$	$(W_2 - W_1)/h$
W_2	W_0	$W_2 - W_0$	$(W_2 - W_0)/h$
W_1	W_0	$W_1 - W_0$	$(W_1 - W_0)/h$

In one case, the energies of three levels above the ground state are 1.06×10^{-14} J, 1.10×10^{-14} J and 3.00×10^{-14}. List the six quanta the system can emit, in order of wavelength. ($h = 6.60 \times 10^{-34}$ J s; $c = 3.00 \times 10^8$ m s^{-1}.)

$_1W_o = W_1 - W_o = 1.06 \times 10^{-14}$ J $h = 6.6 \times 10^{-34}$ J s
$c = 3.00 \times 10^8$ m s^{-1} $\lambda_o = ?$

$_1\nu_o = {}_1W_o/h = 1.06 \times 10^{-14}/6.6 \times 10^{-34}$ Hz

$_1\lambda_o = c/{}_1\nu_o = 3.00 \times 10^8 \times 6.6 \times 10^{-34}/1.06 \times 10^{-14}$ m
$= 1.87 \times 10^{-11}$ m.   [462]

Now take the case of a large number of systems like this. If a lot of energy is contained, many of the particles will be excited. These will have energy W_1, W_2 or W_3; from these levels they will 'fall' and emit quanta. Each quantum will have one of the six allowed energy values.

Thus the spectrum of the energy given out by this sample will be a series of lines. In Fig. 12.11, such a **line spectrum** is plotted (this time using a linear scale); we use the answers to the last question.

The radiation from any simple excited sample has a line spectrum. This is true for all regions of the electromagnetic spectrum.

Line spectra (such as in Fig. 12.11): these appear when the radiation sources (nuclei, atoms, etc.) are separate, as in a gas or vapour composed of single atoms.

Band spectra (see Fig. 12.12): from separate groups of sources, such as the molecules of a polyatomic gas or vapour like CO_2 and CH_4.

Continuous spectra (like that of Fig. 12.4): from close sources, as in solids and liquids.

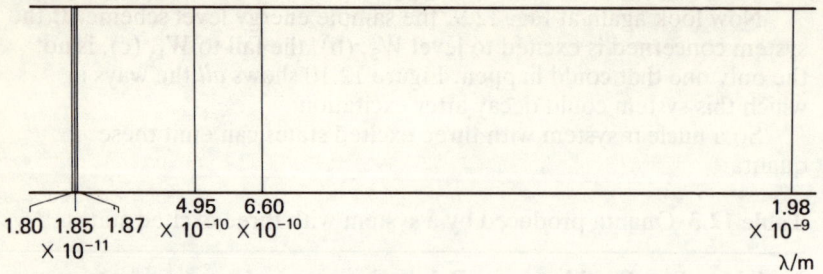

1.80 1.85 1.87   4.95   6.60   1.98
$\times 10^{-11}$ $\times 10^{-10}$ $\times 10^{-10}$ $\times 10^{-9}$

λ/m

Note: This time a linear scale has been used.

Fig. 12.11 The line spectrum of a system with three excited states

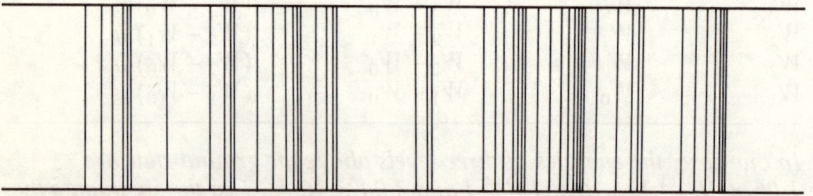

Fig. 12.12 A band spectrum

As Fig. 12.12 shows, energy level schemes can be very complex. This is true even for simple sources, giving simple line spectra.

The main point to make is that **the spectrum from each type of nucleus, atom or molecule is unique. Spectroscopy** is the identification of substances in the source from the spectrum. It is a most important tool in all the sciences. To obtain the most useful results, each source should be in gas or vapour form. It is then excited by raising it to a high temperature or by passing a current through it.

In Fig. 12.13, (a), (b) and (c) are sections of the spectra of three substances A, B and C. (d) shows the spectrum of an unknown substance X. What can you say about X?   [481]

Fig. 12.13 Simple spectroscopy

A second sample Y contains a little A and a lot of B. C is absent. Draw the spectrum of Y. [499]

How is sample Z made up of D, E, F and G? (See Fig. 12.14.) [434]

Fig. 12.14

12.6 Absorption spectra

So far we have discussed **emission spectra**, the spectra given by matter samples previously excited to high energy falling to lower energies. However, we know that matter absorbs energy as well as being able to emit it. Indeed matter *must* absorb energy before it can radiate, otherwise the particles cannot become excited.

The quantum theory can deal with the absorption of energy by a system as well as its radiation. To look at this, we shall use a system whose energy levels (Fig. 12.15(a)) relate to visible quanta rather than gamma-rays. The concepts apply to all types of radiation, however.

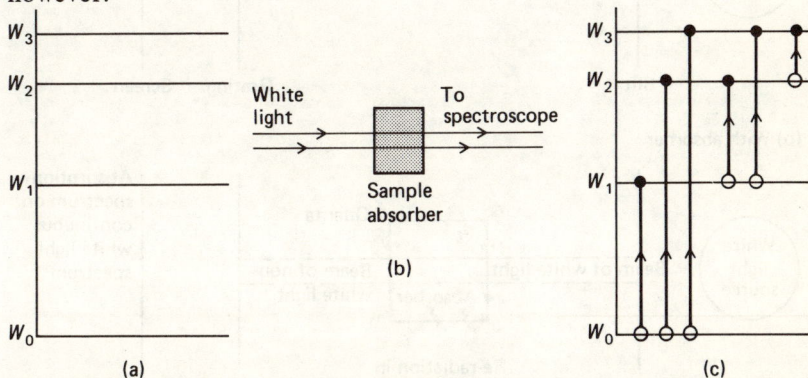

Fig. 12.15 How matter absorbs radiation

In Fig. 12.15(b), white light is shown meeting a sample. Photons will be absorbed from it *only* if they can excite the particles to allowed energy levels. The rest of the photons pass through. In the case shown, photons of the six possible energy values will be absorbed: $W_3 - W_2$, $W_3 - W_1$, $W_3 - W_0$, $W_2 - W_1$, $W_2 - W_0$, and $W_1 - W_0$. (See Fig. 12.15 (c).)

The observed light will no longer be white; its continuous spectrum will have six black lines X at the wavelengths corresponding to the photons absorbed. (See Fig. 12.16.)

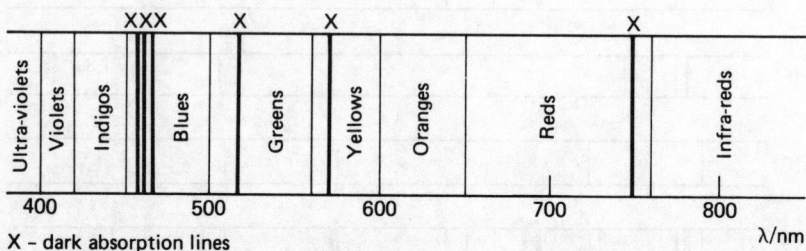

X – dark absorption lines

Fig. 12.16 An absorption spectrum in the visible section

Of course, after each of the many particles is excited, it will tend to 'fall down' the energy level ladder again. In doing so, it will emit the corresponding quanta. However, only a tiny fraction of these quanta will escape in the direction of the original light beam. So the dark lines of the absorption spectrum will not be refilled. (See Fig. 12.17.)

(a) Without absorber

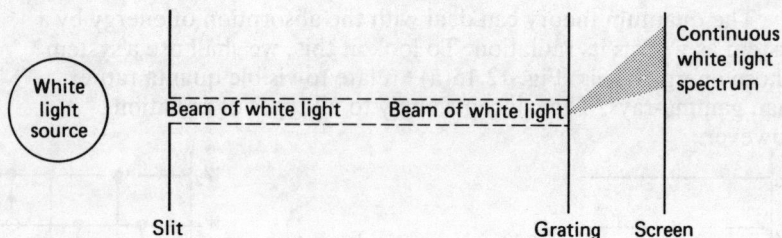

(b) With absorber

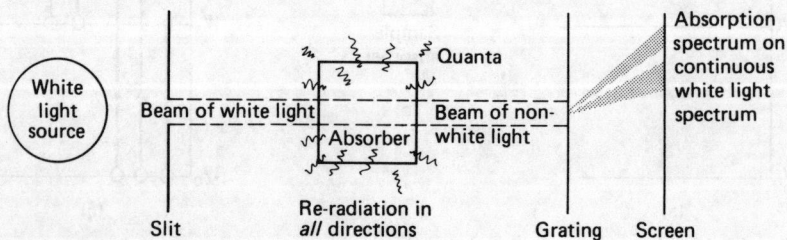

Fig. 12.17 Forming an absorption spectrum

*In an absorption spectrum, a dark line appears at 500 nm. What energy
level change relates to this?*

$\lambda = 500 \times 10^{-9}$ m $h = 6.6 \times 10^{-34}$ J s $c = 3.00 \times 10^8$ m s^{-1} $W = ?$
$W = h\nu$ $c = \nu\lambda \rightarrow W = h\,c/\lambda$
$= 6.6 \times 10^{-34} \times 3.00 \times 10^8/500 \times 10^{-9}$ J $= 3.96 \times 10^{-19}$

You can see that we deal with these problems just as with those on
emission.

*The sodium spectrum shows two very close lines, at 589.0 nm and
589.6 nm. Find the energy level change that relates to each. If these two
changes are $W_2 - W_0$ and $W_1 - W_0$, find the wavelength that relates to
$W_2 - W_1$. In which part of the spectrum would this lie?* [214]

The samples used for absorption spectroscopy are often in the form
of gases at room temperature. As before, solids and liquids tend to
produce continuous absorption spectra, while compound molecules
have band spectra.

Light filters are often made of glass in which is dissolved one or
more compounds able to absorb unwanted wavelengths. (Stained glass
is very much the same.)

The **solar spectrum** (i.e. that of the Sun) is a continuous emission
spectrum. It is continuous because the Sun consists of gases at very
high pressure (in this sense, more like a liquid therefore). Careful
viewing of this spectrum, however, shows many dark absorption lines.
These allow the Sun's low pressure atmosphere to be analysed. An
early triumph of spectroscopy was the discovery of helium; this gas
showed its presence in the solar atmosphere by reason of dark spectral
lines that could not be matched with known patterns.

12.7 Some more questions

1. *Discuss with care how grating and prism spectroscopy differ and how
   they are alike.* [149]
2. *Complete Table 12.4 ($h = 6.6 \times 10^{-34}$ J s; $c = 3.00 \times 10^8$ m s^{-1}).*

Table 12.4

	Quantum wavelength	Quantum frequency	Quantum energy	Spectrum region
(a)	60 nm			[154]
(b)	1500 nm			[137]
(c)		3×10^{20} Hz		[163]
(d)		3×10^{10} Hz		[187]
(e)			1 J	[203]
(f)			10^{-20} J	[237]

3. *Construct an energy level scheme for a system with four excited states. It should be rather like Fig. 12.9. Assign values to the levels such that the line spectrum lies entirely within the microwave region. Use those values to find the wavelengths of the ten allowed quanta; draw the line spectrum to a linear scale.* [378]

4. *Write notes, with sketches, on how you think fluorescence (section 12.2) is explained by the quantum theory.* [408]

5. *What kind of spectrum would you expect for chlorophyll?* [451]

6. *'Hydrogen gas gives the simplest spectrum of all' (section 12.5). Why do you think this is?* [470]

7. *Refer to Fig. 12.14. Draw the line spectrum of a mixture of equal parts of E and G and a lot less F, but no D.* [489]

12.8 Objectives

When you have studied this chapter, you should be able to

(1) describe, explain and compare the dispersion of light by a grating and by a prism;

(2) describe the continuous spectrum of white light;

(3) relate ultra-violet and infra-red radiations to visible light, and say something of their nature;

(4) define ultra-violet fluorescence, giving some of its uses;

(5) describe experiments with microwaves to show that they behave like light;

(6) use the Young's two-slit wavelength relation with any part of the electromagnetic spectrum;

(7) outline the electromagnetic spectrum, showing some knowledge of the sources and nature of its component radiations;

(8) provide a simple account of the quantum theory;

(9) discuss simple energy level schemes and relate them to line emission and absorption spectra;

(10) state how line, band and continuous spectra differ;

(11) describe the characteristics and production of absorption and emission spectra;

(12) state that spectra are characteristic of their sources;

(13) use spectrograms of standards to identify elements in unknown spectrograms.

Chapter 13

Vibration

13.1 Simple harmonic motion (s.h.m.)

In the last three chapters we saw how waves behave. We met some simple pictures of what waves are, but ignored the details.

Here we look at wave motion with more care. How is a wave produced? How does it travel through a medium?

The answer to both questions is – **vibration**. Thus an object vibrates to produce sound; the sound travels through matter by the vibration of particles; a graph of the pressure in the medium has a sine wave form. Both when a sound wave is produced, and as it travels, something vibrates.

Vibration is a regularly repeated to and fro movement. It can be called **oscillation**, or **harmonic motion**. (It is a type of **periodic motion**, or **cyclic motion**.)

So **sound is produced by the vibration of an object; it travels by the vibration of the particles in the medium**.

All waves are produced, and travel, by some form of vibration. In the case of light, the source is vibrating electric charge; the waves themselves are vibrating electric and magnetic fields. (See Fig. 12.7.)

So **light is produced by vibrating electric charge; it travels by the vibration of electric and magnetic fields**.

In practice the changes in a wave can be very complex. Figure 13.1 shows a sound wave from a gently played flute – in fact quite a pure

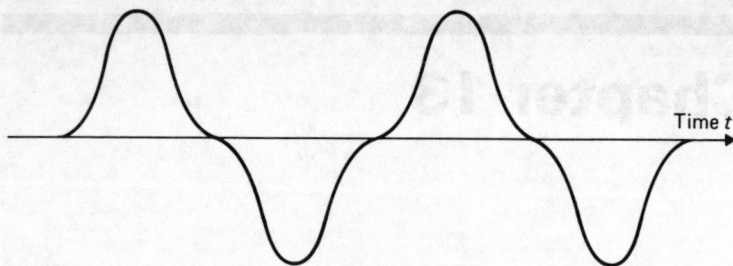

Fig. 13.1 A sound wave from a flute

sound. The sound waves from most sources are far more complex than this; the sound wave you hear from a full orchestra is even more complex still.

In section 11.1 we found that simple wave forms can be added to make complex ones. (This uses the **principle of superposition**.) The reverse is also true.

Any complex vibration is the sum of simple ones, each with its own period, amplitude and phase. (Check with section 11.2 if you need to.)

This concept has great value in many parts of physics. If one knows about simple harmonic motion, one can discuss the most *complex* cases.

Any **simple harmonic motion** (s.h.m.) has three characteristics. Check how they apply to the examples in Fig. 13.2.

The characteristics are in the box below. First we must define a few terms; some we have met before, mainly in section 10.1.

A cycle is a complete to and fro movement.

The displacement is the distance from the centre.

The amplitude is the highest displacement.

The phase of a wave is a measure of how far it has passed through a cycle.

The period is the time for one cycle.

The restoring force is the force trying to return the system to the centre.

A motion is simple harmonic if
 (i) The displacement varies with time like a sine wave.
 (ii) The period does not depend on amplitude.
 (iii) The restoring force is proportional to the displacement.

Look at Figure 13.2(b), for instance; try to picture how the ball rolls to and fro in the bowl. It moves fast at the centre and less and less

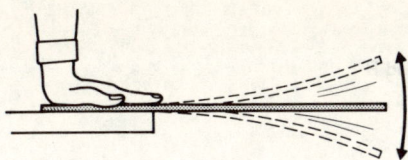

(a) The end of a ruler 'twanged' on the edge of a table

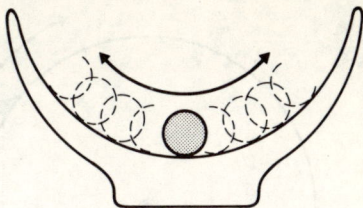

(b) A marble rolling inside a bowl

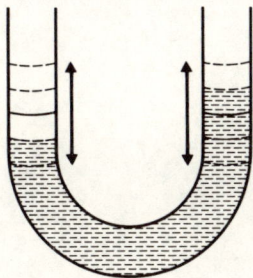

(c) Liquid 'sloshing' to and fro in a U-tube

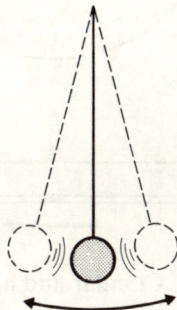

(d) The simple pendulum

Note: These may be strictly s.h.m. only for small amplitudes

Fig. 13.2 Some cases of simple harmonic motion (s.h.m.)

quickly up the sides. A graph of its distance from the centre is like a sine wave. The time for a single to and fro roll is fairly constant (test this yourself!). The force pulling the ball to the centre relates to displacement. (Here the force is equal to the component of the weight parallel to the surface.)

It is helpful to look at the other motions in Fig. 13.2 in the same way.

Before we apply these ideas to given cases in more detail we must relate simple harmonic motion to circular motion.

13.2 Circular and harmonic motions

We discussed circular motion in Chapter 3. It is the motion of an object tracing a circular path at steady speed. Clearly it is a periodic process – the same path is followed at regular time intervals.

Harmonic motion (vibration) – a regularly repeated to and fro movement – is also cyclic.

Can we relate the two?

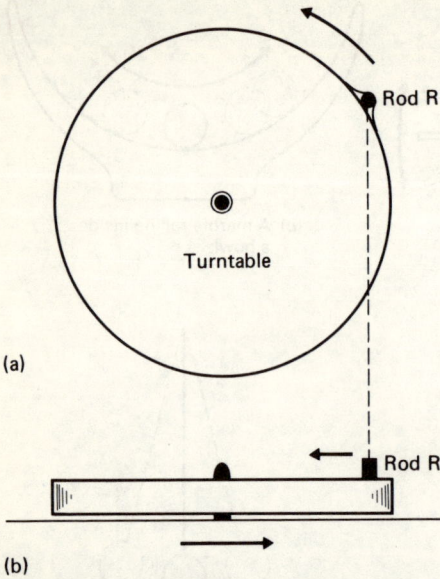

Fig. 13.3 Circular and harmonic motions

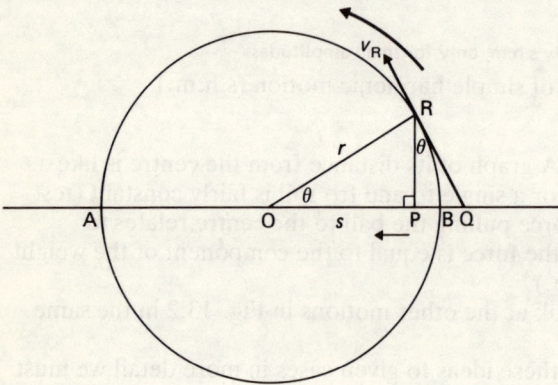

Fig. 13.4 Relating circular and harmonic motions

Figure 13.3 (a) shows the circular motion (like that in Fig. 3.2) of a rod fixed to the rim of a record-player turntable; (b) is a side view of this. From the side the rod seems to move to and fro in a regular way. Does it vibrate with s.h.m.?

Figure 13.4 shows the picture in a more helpful way. R is the rod; P is its projection – or 'shadow' – on the diameter AB. The amplitude of P's motion is r, the circle radius.

How does **displacement** s from O depend on time? Look at triangle ORP.

$$s = r \cos \theta.$$

But $\qquad \theta = \omega t$ $\qquad$ (section 3.1)

So $\qquad s = r \cos \omega t$

A plot of s against t will be like a sine wave.

Thus the motion of P shows the first characteristic of s.h.m. – **the displacement varies with time like a sine wave.**

The second concerns the **period** of the motion – the time for one cycle.

What is the period of P's motion? It is the same as that of R. This is because it takes the same time for R to go round the circle as for P to go from B to A and back to B.

$T = 2 \pi/\omega$ $\qquad$ (see section 3.2)

So **the period of P's motion does not depend on the amplitude** – this is the second characteristic of s.h.m. (We can relate this to the record player turntable discussed in Chapter 3.)

In fact the period depends only on ω. Recall that ω is the angular velocity of an object moving in a circle. That name has no meaning for simple harmonic motion; here we call ω the **pulsatance** of the motion.

The pulsatance ω of an s.h.m. is 2 π divided by the period.

$\omega = 2 \pi/T$ $\qquad$ Unit: hertz, Hz

As $\qquad T = 1/v$ $\qquad$ (v is frequency; see section 10.1)

we can also write $\qquad \omega = 2\pi v$

Thirdly, we look at the **restoring force** on P.
What is the acceleration a_R of R (the rod in Fig. 13.4) towards O? It is the centripetal acceleration . (See section 3.3.)

$$a_R = v^2/r \qquad (v \text{ is R's velocity}).$$

But $\qquad \omega = v/r$ $\qquad$ (section 3.2).
So $\qquad a_R = \omega^2 r$

What is the acceleration a of P (the rod's projection in Fig. 13.4) towards O? It is the component along AB of R's acceleration.

$$a = a_R \cos \theta$$
$$= \omega^2 r \cos \theta \qquad (\text{towards O}).$$

But $\qquad s = r \cos \theta$

So $\qquad a = \omega^2 r \qquad\qquad$ (towards O)

or $\qquad a = -\omega^2 s \qquad\qquad$ (in the same direction as s).

In other words – the deceleration of P is proportional to displacement. As deceleration clearly relates to restoring force (from $F = ma$), we meet the third characteristic of s.h.m. **The restoring force is proportional to the displacement**. Do not forget that s and a are vectors – they have opposite signs as they are in opposite directions.

The record-player turntable in Fig. 13.3(a) is 300 mm in diameter; it rotates at 33 rev min⁻¹ Viewed as in (b) the rod moves with s.h.m. Find (a) the frequency; (b) the period; (c) the pulsatance; (d) the acceleration, when the displacement is (i) maximum, (ii) minimum.

(a) $\nu = 33$ rev min^{-1} = 33/60Hz = 0.55 Hz;

(b) $T = 1/\nu = 60/33$ s = 1.8 s;

(c) $\omega = 2\pi\nu = 2\pi \times 33/60$ Hz = 3.5 Hz;

(d) (i) $a = -\omega^2 s = -(4\pi^2 \times 33^2/60^2)\, 0.15$ m s^{-2}
$\qquad\qquad$ = 1.79 m s^{-2} (when s is maximum),

$\qquad$ (ii) $a = -\omega^2 s = 0$ m s^{-2} (when s is zero).

A tuning fork vibrates at 512 Hz. What is the highest acceleration of the tip, if the amplitude is 0.25 mm? $\qquad$ [143]

What have we proved in this section? We have shown that the projection on the diameter of an object moving in a circle, moves with s.h.m.

The reverse is also true. **Any s.h.m. can be viewed in terms of circular motion**. This is the case, too, with any sine wave form, as that is also an aspect of s.h.m. **Any wave form can be viewed in terms of circular motion**. The approach is called the **phasor** view. In this section the phasor – a rotating vector – has been the line OR in Fig. 13.4.

Any simple harmonic motion and any simple sine wave form can be expressed as a phasor (a rotating vector).

13.3 Examples of s.h.m.

Look at Fig. 13.5. A load vibrates without friction at the end of a spring.

What happens if the stationary object is pulled to the right through a distance s? A tension will appear in the spring, trying to pull it back. From Hooke's law (section 6.3), this restoring force F is proportional to s. This will be true for any value of s, positive or negative, as long as the proportional limit of the spring is not passed.

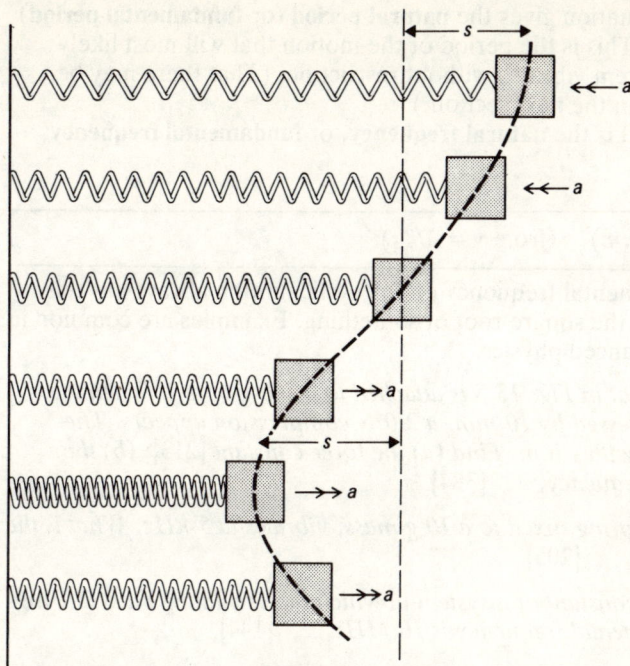

Fig. 13.5 s.h.m. of an object on a spring

So the object, when released, will experience a **restoring force which is proportional to the displacement**.

We can write $F = -ks$.

Here k is the **spring constant** (or **force constant**).

$k = -F/s$ Unit: newton per metre Nm^{-1}

The constant of a spring (or any other Hooke's law system) is the force needed to produce unit change of length.

Now $-ks = ma$ (as $F = ma$)

or $a = -(k/m)s$

But $\omega = \sqrt{k/m}$ (as $a = -\omega^2 s$).

So $T = 2\pi\sqrt{m/k}$ (as $T = 2\pi/\omega$)

The period of vibration of this system does not depend on amplitude.

So, if the object in Fig. 13.5 is pushed or pulled from the rest position and released, it will vibrate to and fro with s.h.m. A graph of its position against time will be a sine wave.

The last equation gives the **natural period** (or **fundamental period**) of the motion. This is the period of the motion that will most likely occur if the system vibrates without restriction. (That there may be others, we see in the next section.)

More useful is the **natural frequency**, or **fundamental frequency**, v_1.

$v_1 = (1/2\pi)\sqrt{(k/m)}$ (from $v = 1/T_1$).

The fundamental frequency of any cyclic motion always has this form – $1/2 \pi \times$ the square root of something. Examples are common in all parts of advanced physics.

A spring like that in Fig. 13.5 is attached to a 500 g object. When the spring is compressed by 10 mm, a 2.0 N compression appears. The object vibrates with s.h.m. Find (a) the force constant [213]; (b) the fundamental frequency.   [384]

Another such spring, fixed to a 10 g mass, vibrates at 5 kHz. What is the force constant?   [205]

Find the force constant of a system in which the oscillating mass is 1 mg and the fundamental frequency is 10 MHz.   [144]

13.4   Displacement, velocity and acceleration

For an object in s.h.m., how do these three measures relate?

In section 1.2 we saw that an object's **velocity is the displacement in unit time**. That was good enough for constant velocity work; now, however, the velocity is not constant. A better definition is – **velocity is the rate of change of position with time**. It is given by the slope of the displacement–time graph (from section 1.1).

To find the object's velocity at any time, we find the slope of the displacement–time curve at that time. If the slope is large, the velocity is high; if the slope is small, the velocity is low.

In Fig. 13.6(a) the displacement–time curve for s.h.m. is shown again. Below it, (b), are some points marked on velocity–time axes. Velocity is zero when the s–t curve is flat; it is highest when that curve is steepest.

Can we join those points in a smooth curve? If we do, we get the waveform shown in Fig. 13.6(c). If you suspect that this is a sine wave form, you would be right. (We prove this shortly.)

The velocity–time curve for s.h.m. is like a sine wave.

We use the same approach to find how the object's acceleration depends on time. Again we adapt the ideas in Chapter 1.

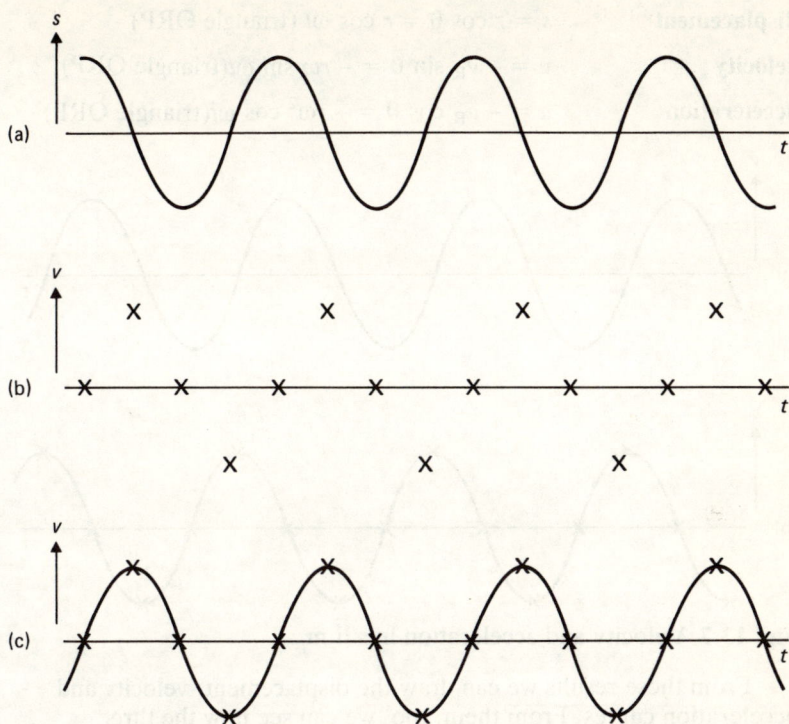

Fig. 13.6 Displacement and velocity in s.h.m.

Acceleration is the rate of change of velocity with time. To find an object's acceleration at any time, we find the slope of the velocity–time curve at that time.

Figure 13.7(a) gives the velocity–time curve we have just found. In (b) is the corresponding acceleration–time curve. This, too, seems like a sine wave (and this too we shall shortly prove).

What can we conclude from these curves?

Conditions for s.h.m.:

An object moves simply harmonically if these conditions apply:
 (i) Its acceleration is proportional to its displacement, but in the reverse direction.
 (ii) Its velocity is highest at zero displacement, and zero at highest displacement.
(iii) Its displacement, velocity and acceleration all vary with time like a sine wave.

The same results follow from Fig. 13.4; we use right angled triangles to give:

154

displacement	$s = r \cos \theta = r \cos \omega t$ (triangle ORP)
velocity	$v = - v_R \sin \theta = - r\omega \sin \omega t$ (triangle QRP)
acceleration	$a = - a_R \cos \theta = - r\omega^2 \cos \omega t$ (triangle ORP)

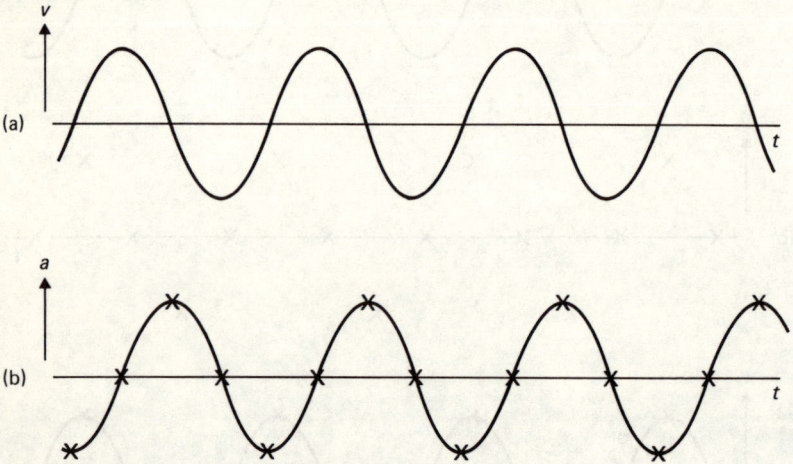

Fig. 13.7 Velocity and acceleration in s.h.m.

From these results we can draw the displacement, velocity and acceleration curves. From them, too, we can see how the three measures vary with time between highest and lowest values. The same results can be achieved using differential calculus.
(If you can do differential calculus, here is the approach.

displacement	$s = r \cos \omega t$
velocity	$v = ds/dt = - r\omega \sin \omega t$
acceleration	$a = dv/dt = - r\omega^2 \cos \omega t$

The results, and the conclusions, are the same.)

13.5 Resonance

We saw in section 13.3 that if a system can vibrate with s.h.m., it will do so at a particular frequency; this is called its **fundamental frequency**, or natural frequency. In practice, however, certain other frequencies are possible.

We can describe this with the ways an object vibrates to produce sound.

Consider a stretched guitar string. If plucked, it will tend to vibrate as in Fig. 13.8(a). Sound waves will result; their frequency depends on the string's length, tension and density.

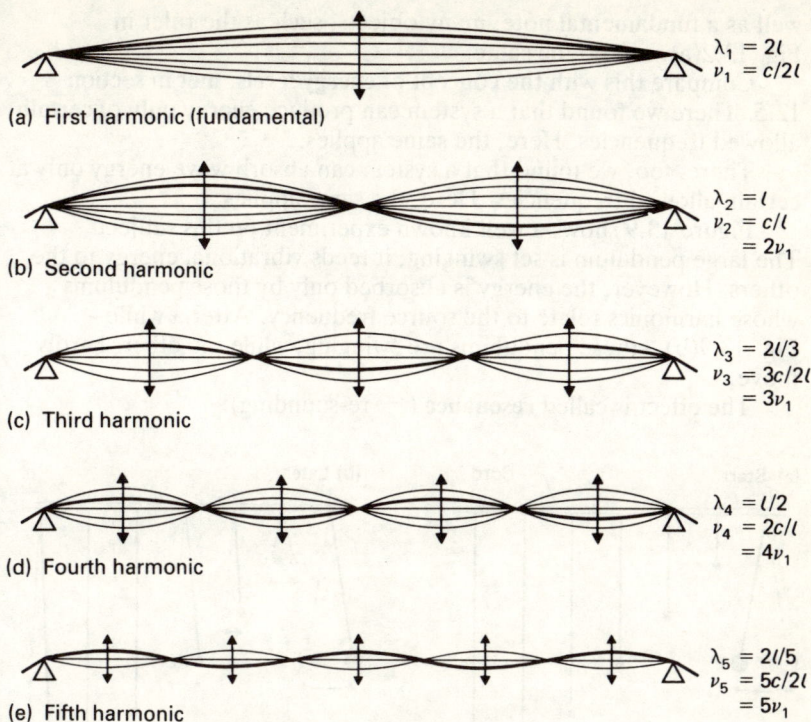

$\lambda_1 = 2l$
$\nu_1 = c/2l$

(a) First harmonic (fundamental)

$\lambda_2 = l$
$\nu_2 = c/l$
$= 2\nu_1$

(b) Second harmonic

$\lambda_3 = 2l/3$
$\nu_3 = 3c/2l$
$= 3\nu_1$

(c) Third harmonic

$\lambda_4 = l/2$
$\nu_4 = 2c/l$
$= 4\nu_1$

(d) Fourth harmonic

$\lambda_5 = 2l/5$
$\nu_5 = 5c/2l$
$= 5\nu_1$

(e) Fifth harmonic

Fig. 13.8 The first few harmonics of vibration of a spring

The natural period, T_1, the natural frequency, ν_1, and the natural wavelength, λ_1, of the fundamental vibration of the string, describe the sound produced.

In the case shown in Fig. 13.8(a), the wavelength of the sound is twice the length l of the string.

$$\lambda_1 = 2l$$
Therefore $\nu_1 = c/2l$ (from $\nu = c/\lambda$, c being the speed)
and $T_1 = 2l/c$

However, the string *could* also vibrate in another **mode**; this is shown in Fig. 13.8(b). This is not as likely as the first – but it can, and does, happen.

This time a sound is produced with twice the frequency of the first. In music this is called the first **overtone**; in physics we call it the second **harmonic**. (The fundamental mode is the first harmonic.)

A whole series of vibrations can be produced by a string in this way. This first few are shown in Fig. 13.8.

Most vibrating systems can show a whole series of frequencies (or harmonics) like this. Any musical instrument produces overtones as well as a fundamental note; many objects, such as the ruler in Fig. 13.2(a), can do the same.

Compare this with the concept of **energy levels**, met in section 12.5. There we found that a system can produce energy only of certain allowed frequencies. Here, the same applies.

There, too, we found that a system can absorb wave energy only at certain allowed frequencies. Here, the same applies.

Figure 13.9 shows a well known experiment on this subject. The large pendulum is set swinging; it feeds vibrational energy to the others. However, the energy is absorbed only by those pendulums whose harmonics relate to the source frequency. After a while – Fig. 13.9(b) – these pendulums are swinging, while the others hardly move.

The effect is called **resonance** (= re-sounding).

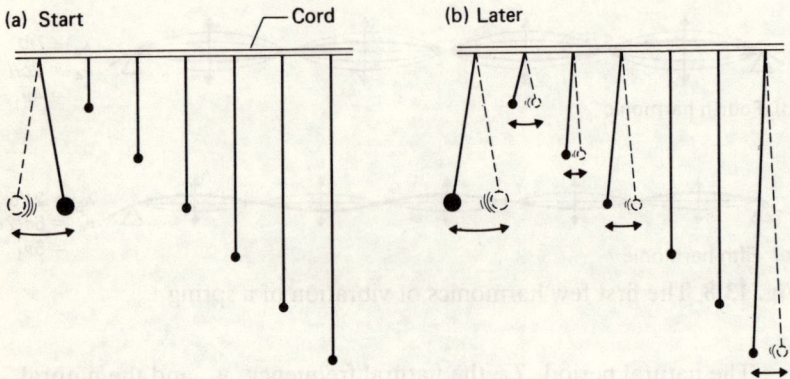

Fig. 13.9 Resonance in a set of pendulums

Resonance is when a system starts to vibrate at one of its harmonics when fed with vibration energy at that frequency.

There are many cases in practice.

Sing a note near a piano – some of the strings will start to vibrate 'in sympathy'. The strings that resonate have harmonics that relate to the source frequency. When you blow over the top of an empty bottle, the air inside may resonate, to produce a 'hoot'.

At certain engine speeds a car body may vibrate strongly. In cases like this damage could occur. Damage may be caused to suspension bridges resonating in gusty winds and to buildings resonating in an earthquake.

Certain electric circuits have resonant frequencies; they will pass current well at these frequencies but not at others. Radio and TV tuners work like this.

A spring has force constant 100 Nm^{-1}. One end is fixed; the other is *attached to a 1 kg mass. At what earthquake frequency would the system vibrate most?*

$k = 100 \text{ Nm}^{-1}$ $m = 1.0 \text{ kg}$ $v_1 = ?$

$v_1 = (1/2\pi) \sqrt{k/m} = 1/2\pi \sqrt{100/1.0} \text{ Hz} = 1.6 \text{ Hz}$

It is important to note that while every object or system has a natural frequency, it can resonate when fed with energy only if it is *free* to vibrate. If a vibration is very hard to start or to maintain, we say the object is heavily **damped**. Damping , in effect, measures resistance to vibration; it is important in many fields. A typical example of damping is the use of shock absorbers in a car which reduces the vibration of the car on an uneven road.

Resonance can occur only in lightly damped systems This is the result of the applied frequency being very close to the natural frequency of the system. If the applied frequency is not very close the damping is high.

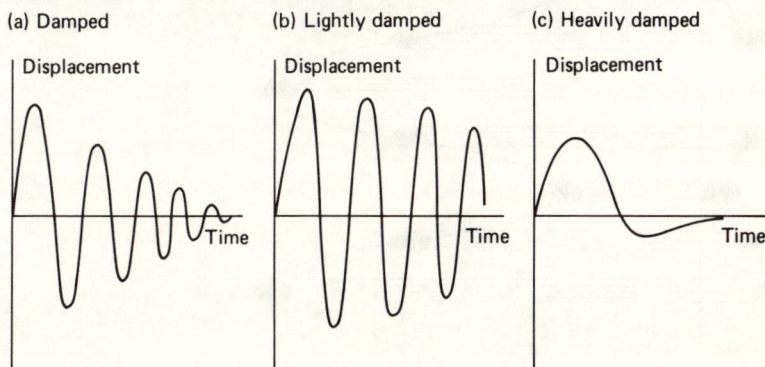

Fig. 13.10 The vibration of damped systems

13.6 Resonance and absorption spectra

We noted above that the series of harmonics of a system able to vibrate compares to the system of energy levels allowed by quantum theory.

The reason is that to a large extent, **the forces between the particles of matter follow Hooke's law.**

Figure 13.11 shows a simple case – the diatomic (two atom) particle of hydrogen chloride gas, HCl. Here the light H-particle and the massive Cl-particle are tightly bound to each other; the forces concerned give a certain equilibrium distance of separation.

The HCl particle has energy – it can move in many ways. Our concern here is how it vibrates about its centre. Bearing in mind that

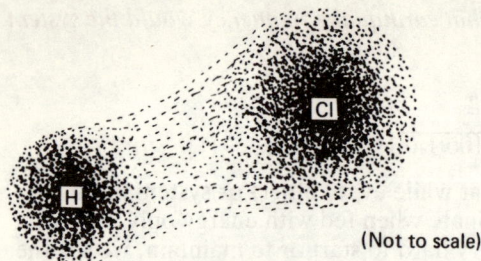

Fig. 13.11 A lone HCl particle

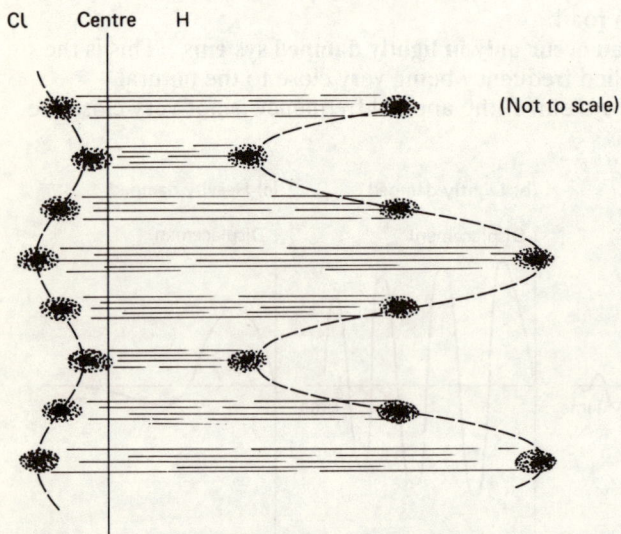

Fig. 13.12 Vibration of a diatomic particle

the mass of the Cl part is many times that of the H part, it will vibrate with s.h.m. as in Fig. 13.12. The H atom vibrates with large amplitude; the massive Cl atom has a very small amplitude – we can think of it as static.

The effect is almost the same as that drawn in Fig. 13.5; that showed how a mass at the end of a spring vibrates. In this case the force constant can be determined if the equilibrium vibration frequency is known. The force constant can then be used to determine the equilibrium distance (*interatomic* distance) between the atoms, and the forces between the atoms (interatomic forces).

We can find the fundamental frequency of a diatomic particle in the same way as we found that for the mass on a spring. To allow for the fact that this time 'both ends of the spring' are moving, the **'reduced mass'** is used to describe the system.

When an HCl particle vibrates, the fundamental frequency is 8.7×10^{13} Hz. The reduced mass is 1.6×10^{-27} kg. Find the force constant.

$v_1 = 8.7 \times 10^{13}$ Hz $m = 1.6 \times 10^{-27}$ kg $k = ?$

$v_1 = 1/2\pi \sqrt{(k/m)}$

$\rightarrow k = 4\pi^2 v^1 m$
$= 4\pi^2 (8.7 \times 10^{13})^2 \times 1.6 \times 10^{-27}$ Nm^{-1} = 478 Nm^{-1}

13.7 Some more questions

1. Refer to Fig. 13.2 (c). For the fundamental vibration $T_1 = 2\pi \sqrt{1/g}$. Here 1 is the total liquid length; g is the acceleration of free fall. In a given case, $1 = 400$ mm. Find v_1. [146]
2. Figure 13.13 extends Fig. 13.2 (a); it shows the fundamental and the next two harmonics. A given steel ruler is 300 mm long. The speed of sound in it is 6 km s^{-1}. Find the frequencies of the three harmonics shown. [388]

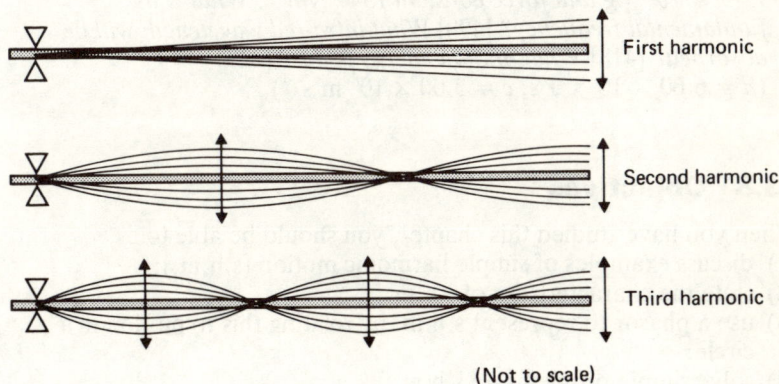

First harmonic

Second harmonic

Third harmonic

(Not to scale)

Fig. 13.13

3. Assume that the ruler in question 2 vibrates only as shown; the amplitudes of the three harmonics are in the ratio 3 : 2 : 1. Draw the three wave forms to scale on axes below each other; use the principle of superposition to find the wave form produced by the ruler. Take as time interval two fundamental periods. [156]
4. At a certain port the tidal flow is simple harmonic; the period is 12.5 h; the tidal range is 15 m. Find (a) the frequency, and (b) the pulsatance of the motion; also find (c) the acceleration of the water surface when the displacement is (i) maximum and, (ii) minimum. [456]
5. An object moves with s.h.m. with a period of 2.00 s. Its highest acceleration is 9.87 m s^{-2}. What is its range of vibration? [475]

160

6. *Figure 13.14 shows a piston attached to a flywheel. Discuss how the motions of piston head and flywheel relate.* [491]

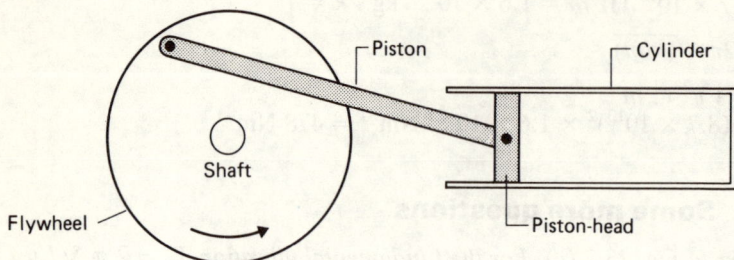

Fig. 13.14

7. *A mass attached to a spring vibrates at 5 kHz. What are the period and pulsatance of its motion?* [482] *If the mass is 110 g, what is the force constant of the spring?* [382]

8. *The nitrogen monoxide, NO, particle has reduced mass 12.4×10^{-27} g and force constant 1540 Nm^{-1}. What is its fundamental frequency?* [498] *What infra-red wavelength will be absorbed?* [418] *What photon energy is this?* [347] ($h = 6.60 \times 10^{-34}$ J s; $c = 3.00 \times 10^8$ m s^{-1})

13.8 Objectives

When you have studied this chapter, you should be able to
(1) discuss examples of simple harmonic motion (s.h.m.);
(2) state the characteristics of s.h.m.;
(3) use a phasor to represent s.h.m. by relating this to motion in a circle;
(4) solve simple problems on s.h.m.;
(5) find the fundamental frequency of a mass which vibrates at the end of a spring;
(6) derive, discuss and relate waveforms showing the displacement, velocity and acceleration of an object moving with s.h.m.;
(7) define resonance and list examples from different parts of physics;
(8) note the damage that can occur in systems as a result of resonance;
(9) find the fundamental and higher harmonic frequencies of a system given its modes of vibration;
(10) discuss the vibration of a diatomic particle and perform simple calculations on its fundamental frequency;
(11) relate the absorption line spectrum of such a particle to the possible harmonics of its vibration.

Part 4

Electricity and magnetism: energy and electric charge

Chapter 14

Electric charge and current

14.1 Electric charge and its flow

Electric **charge** (section 12.5) is a major concept. It is central to all work in the last part of this book. Here are the points we now need to know about it; they appear in Level 1.

(i) Tests show that two **types of charge** exist. They are called negative and positive.

(ii) **Like charges repel each other; unlike charges attract each other**.

(iii) Matter is normally **neutral** – there is no net charge as the positive and negative particles balance.

(iv) Charge can be transferred from place to place by various methods. (All involve energy.) Normally this is by the movement of negative electrons. Balance is then lost. At one place there is an excess of negative charge; elsewhere there is less than normal. The charge of the two regions are negative and positive respectively.

(v) Media through which charge can pass with ease are **conductors**; media through which very little charge can pass are **insulators**. Metals are good conductors; dry air, pure water and most plastics are insulators.

There is an *electric field* around a charge – an electric field is a region of space in which there are electric forces (compare the treatment of gravitational fields in section 2.2).

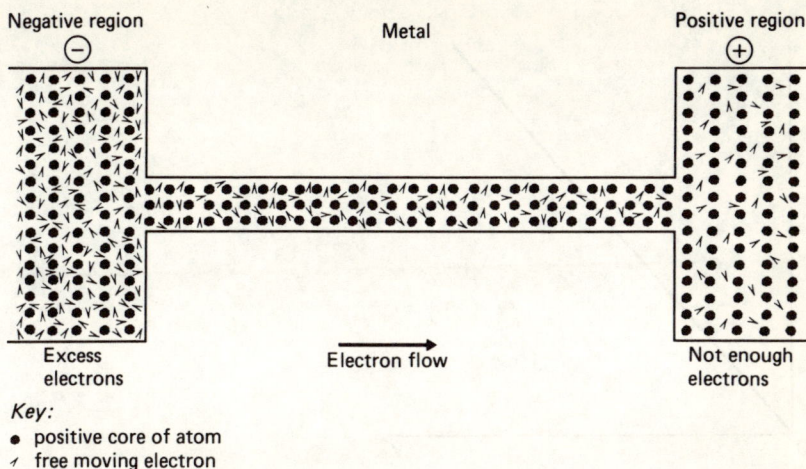

Negative region ⊖   Metal   Positive region ⊕

Excess electrons   Electron flow   Not enough electrons

Key:
• positive core of atom
↗ free moving electron

Fig. 14.1 Conduction of charge

Figure 14.1 shows a length of metal joining positive and negative regions. The free electrons in the metal normally move at random (like the particles of a gas). Now, however, they are forced to drift towards the positive region.

As electrons enter the positive region, more enter the wire at the other end. Unless the two regions are being 'recharged' (using energy), they will lose their charge and the flow to the right will cease.

This drift of electrons forms a **current**.

A current is a flow of charge.

The size of the current depends on the force on the charges. This, in turn, depends on the excess of negative charge at the negative region and the deficiency at the positive region. **Electric potential difference** (p.d.) measures this.

Potential difference (p.d.) measures the difference in charge level between two points.

Ohm's law relates current, I, and potential difference, V, in certain special cases.

The current through a metal sample at constant temperature is proportional to the potential difference between its ends.

$$I \propto V.$$

Thus, if a substance follows Ohm's law, a current/potential difference graph will be like that of Fig. 14.2.

Such a straight-line graph is not usual. Only metals and a few other conductors follow Ohm's law. They are called '**ohmic conductors**'.

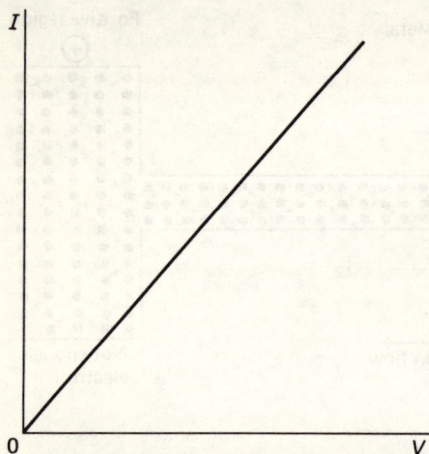

Fig. 14.2 Characteristic of an ohmic conductor

Conductors which do not follow Ohm's law are called non-ohmic.

We return to this in the next section.

From Ohm's law $I \propto V$
which gives $I = kV$

The constant k relates to the other factor on which the current depends.

This is the sample's **resistance R** to the flow of charge.

The resistance of a sample measures how much it opposes charge flow.

If resistance is low, a certain potential difference will cause a large current. If resistance is high, the same potential difference will cause only a small current.

In fact, **the current I in a conductor is the ratio of the potential difference V between its ends to its resistance R.**

$$I = V/R$$

We often use this in the form $V = IR$.

In words: **the potential difference between the ends of a conductor is the product of the current in it and its resistance.**

It is common to extend Ohm's law, and this equation, to any ohmic element of a circuit (section 14.2).

What are the **units** of potential difference, current and resistance?

This unit of current (I) is the ampere, A. We often shorten the name to 'amp'.

The unit of potential difference *(V)* **is the volt, V.**

The unit of resistance *(R)* **is the ohm, Ω.** (Ω is the capital Greek O, omega.)

250 V is applied to a 5 Ω motor. What is the current?

$V = 250$ V $R = 5$ Ω $I = ?$
$I = V/R = 250/5$ A $= 50$ A.

In a circuit a 5 kΩ resistance passes 20 mA. What is the potential difference between its ends?   [284]

What resistance would pass 13 A when 260 V is applied?   [385]

What *is* resistance? Opposition to charge flow is caused as the charge carriers collide with each other and with other particles (such as the 'cores' in Fig. 14.1). The resistance of a sample depends on four factors.

Resistance depends on the substance. For various reasons substances differ in **conductivity**.

Resistance depends on sample length. The longer the wire, the harder it is for particles to pass through.

Resistance depends on the cross-section area. The smaller this is, the harder it is for particles to pass through.

Resistance depends on sample temperature. At higher temperatures, random thermal motion is greater; thus collisions are more frequent. (*Note*, however, that the resistance of some classes of substance falls with temperature rise.)

14.2   Simple direct current circuits

Our concern here is with **circuits** through which a continuous current passes in one direction.

A direct current (d.c.) is a charge flow in one direction. It is normal to take d.c. to be fairly constant except when it is switched on and off.

A d.c. circuit is a source of p.d. and a complete conducting path. Then a current will flow. In practice the circuit will also include a **switch** for simple current control; it will also include one or more other **elements** designed for control or to convert the energy supplied to some other form. Here are some elements: fuse, meter, lamp, motor, fire, electromagnet and electroplating system.

For the moment, however, we are concerned only with the resistance of elements. Also we can ignore the resistance of the connecting wires, for this is usually very low in comparison.

In practice electric circuits can be quite complex. Even a 'simple' outline of a car's lighting system is complex – Fig. 14.3.

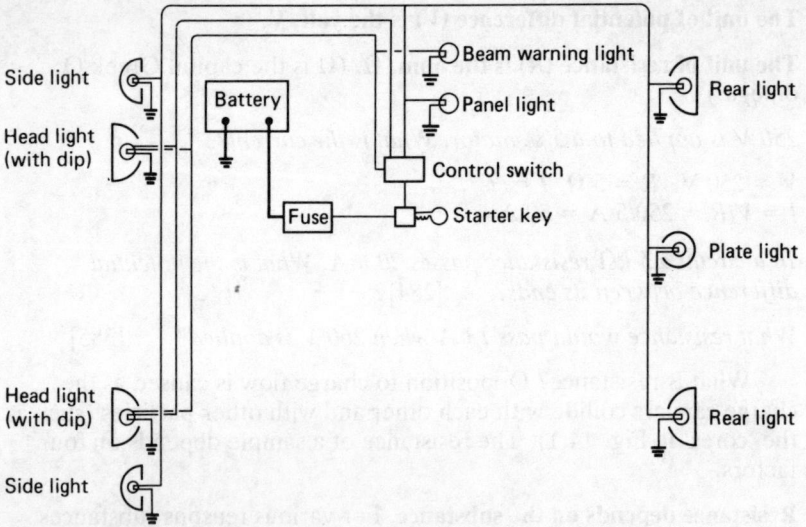

Fig. 14.3 A car lighting system (simplified)

The actual details of how the wires are laid out and so on do not matter in circuit design and analysis. So it is normal to draw simple **circuit diagrams**. Like maps of transport systems, these show only the parts in order and how they join. They are quite enough for most purposes. Figure 14.4 is the circuit diagram of the car lighting system. This may still look quite complex – but it is a lot simpler than before!

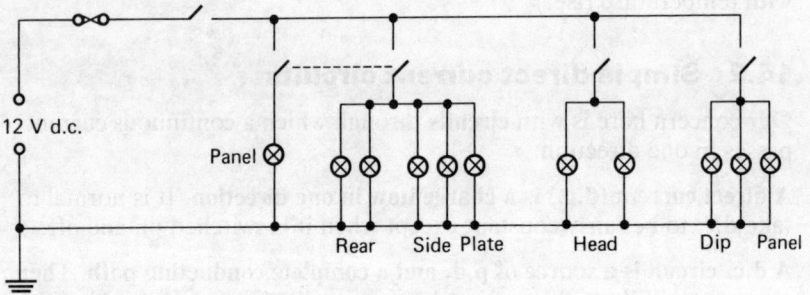

Fig. 14.4 A car lighting system circuit (simplified)

Standard **symbols** are used for all circuit elements. Figure 14.5 gives the ones we need to know now.

When you draw circuits, you should always use the standard symbols. You should also gain practice in building circuits from such diagrams.

In the rest of this section we discuss a number of simple d.c.

circuits. In each case the diagrams are given. Try to build at least
some; that will provide useful practice.

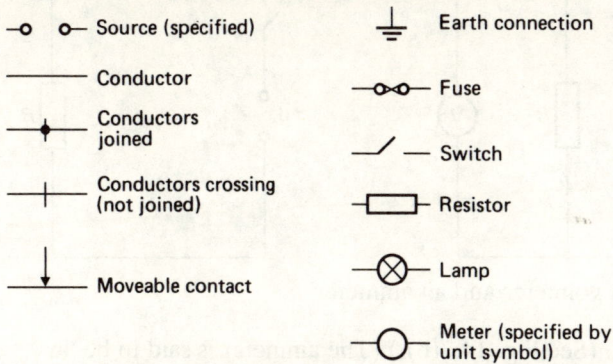

−o o− Source (specified)

───── Conductor

Conductors joined

Conductors crossing (not joined)

Moveable contact

Earth connection

−o⚬o− Fuse

Switch

Resistor

Lamp

Meter (specified by unit symbol)

Fig. 14.5 Basic circuit symbols

Figure 14.6 shows the simplest circuit of all. A direct source feeds
a potential difference V to a loop of resistance R.

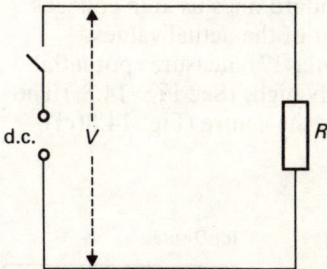

Fig. 14.6 The basic d.c. circuit

*If the p.d. is 9.0 V and the resistance is 36 Ω, what is the
current?*   [233]

14.3 Electric measurements

Meters are used to make electric measurements. We measure potential
differences with a potential difference meter; the common name is the
voltmeter. To find the potential difference between two points – such
as the two ends of the resistor in Fig. 14.6 – the meter is connected
between the points. (See Fig. 14.7(a).) (The voltmeter is said to be 'in
parallel' with R.)

We measure current with a current meter; this is called an
ammeter (= amp-meter). To find the current in an element – like the
resistor in Fig. 14.6 – the meter is connected so that the charge flows

(a)

(b)

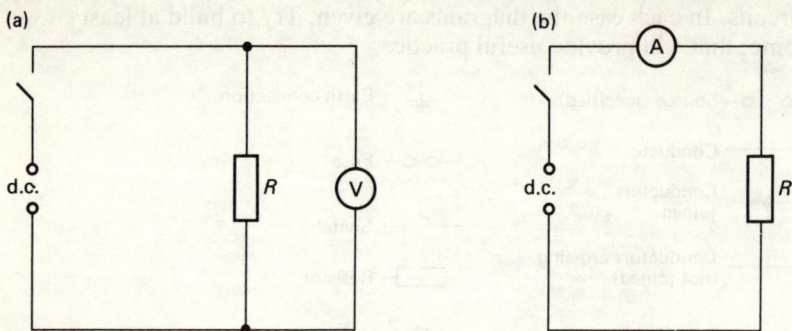

Fig. 14.7 Using a volmeter and an ammeter

through it as well. (See Fig. 14.7(b).) (The ammeter is said to be 'in series' with R.)

Clearly no meter should so affect the circuit that it changes what it measures. In practice, this aim cannot be met – meters need energy to make a reading and must draw this from the circuit under test. Ammeters have very low resistance, and voltmeters have very high resistance to reduce this problem. The standard ones used in colleges can be relied upon to read within 1 per cent of the actual values.

The **oscilloscope** (more on this in Chapter 17) measures potential difference well, as its resistance is extremely high. (See Fig. 14.8.) The distance the spot moves on the screen from the centre (Fig. 14.8(c)) relates to the potential difference applied.

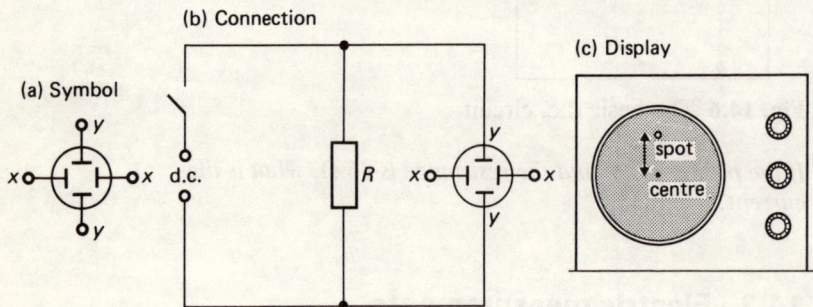

Fig. 14.8 The oscilloscope used to measure p.d.

In section 14.5 we discuss the **potentiometer**; this is even better than the oscilloscope for measuring potential difference.

How can we measure **resistance**? Recall that for any element

$R = V/I$.

R can be found by measuring the potential difference V between the ends of the element, and the current I in it; R is the ratio of these two values.

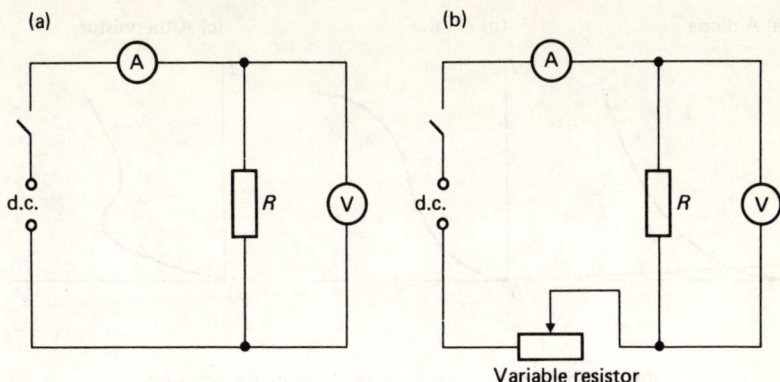

Fig. 14.9 Circuits to measure resistance

A circuit for this is given in Fig. 14.9(a). Use of a variable resistor, Figure 14.9(b), allows us to take a whole set of readings; we can then obtain a mean value of *R*.

This approach can succeed only for ohmic elements (section 14.1). Only these give straight line current/p.d. graphs (like Fig. 14.2); only these have resistances which do not vary with potential difference.

The circuit of Fig. 14.9 (b) can therefore be used to find whether an element is ohmic or not. The current/potential difference graph obtained is the **characteristic** of the element.

The characteristic of a circuit element is the graph that relates the current in it to the potential difference between its ends.

For some non-ohmic characteristics, see Fig. 14.10.

In none of these cases can we speak of the resistance of the element as such – *R* can be defined as *V/I* only at given values of *V* in each case.

We can also use an **ohmmeter** to measure the resistance of an ohmic element. It is in fact an ammeter with its own source. Thus, in effect, the meter finds the value *V/I*. (See Fig. 14.11.)

Multimeters are very useful (Fig. 14.12). A single display is set in a box with switches to allow the user to select the measure to be made. The meter must of course also be connected correctly.

14.4 Series and parallel elements

The phrases 'series connection' and 'parallel connection' were mentioned above. What do they mean?

Elements are in series if the current passes through each in turn.

Elements are in parallel if the current splits to pass through them.

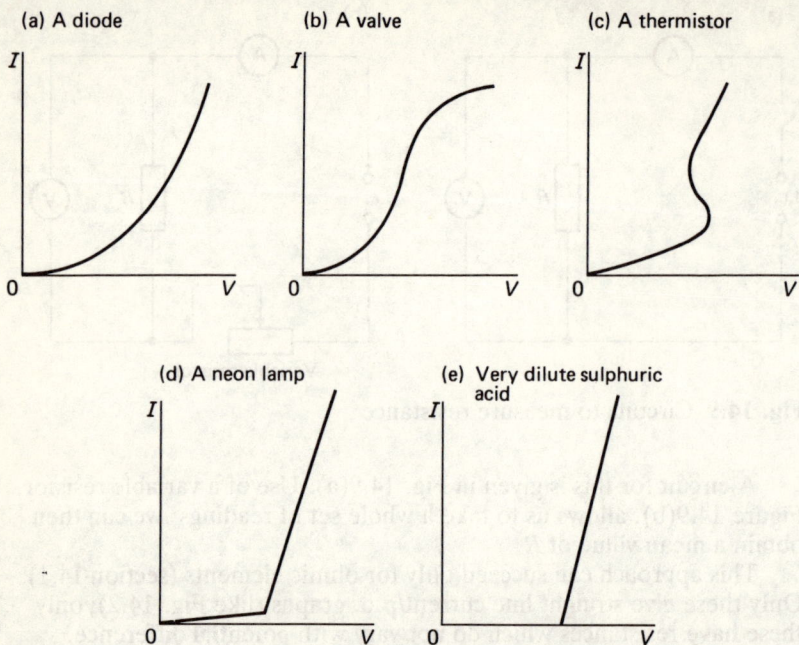

(a) A diode

(b) A valve

(c) A thermistor

(d) A neon lamp

(e) Very dilute sulphuric acid

Fig. 14.10 Characteristics of some non-ohmic elements

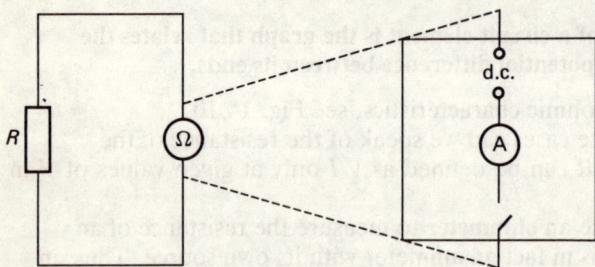

Fig. 14.11 Using an ohmmeter

Figure 14.13 shows three resistors, connected (a) in series, and (b) in parallel.

Tests on **series** circuits show these points.

In a series circuit
(i) the current is the same at all points
(ii) the sum of the potential differences across the elements is the potential difference supplied.

The resistance of the elements in series is the sum of their resistance.

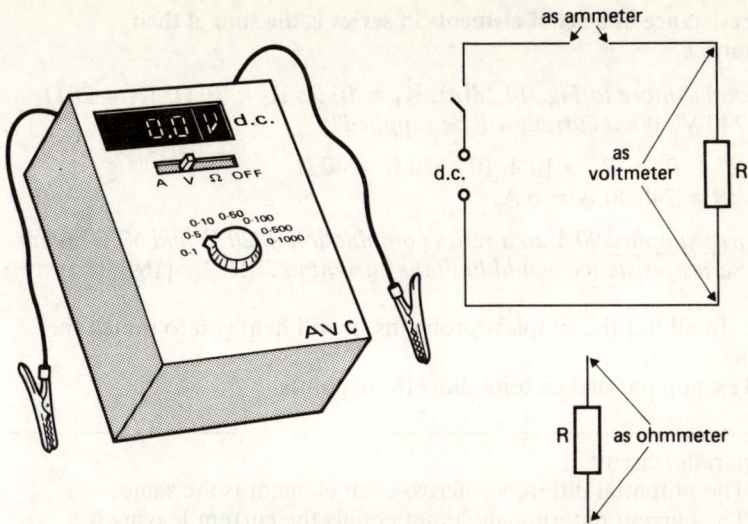

Fig. 14.12 A d.c. multimeter and its use

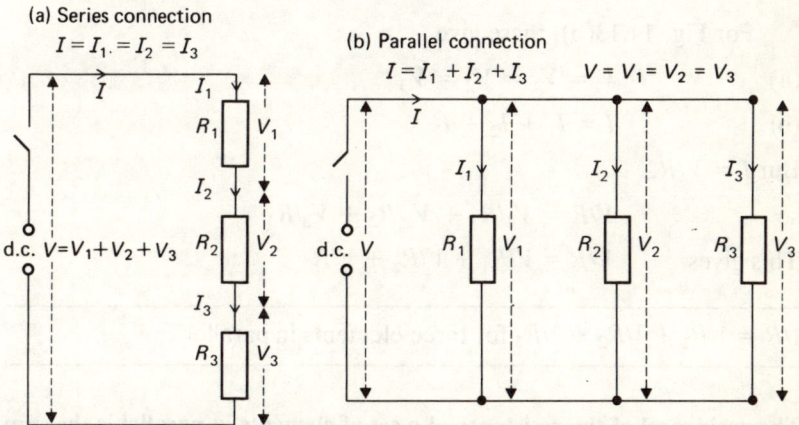

Fig. 14.13

Refer to Fig. 14.13(a) again.

$$\text{(a) } I = I_1 = I_2 = I_3$$

$$\text{(b) } V = V_1 + V_2 + V_3$$

But	$V = IR$ (from Ohm's Law; section 14.1)
So	$IR = I_1R_1 + I_2R_2 + I_3R_3$
This gives	$IR = IR_1 + IR_2 + IR_3$

$R = R_1 + R_2 + R_3$ for three elements in series

The resistance of a set of elements in series is the sum of their resistances.

Refer once more to Fig. 14.13(a). $R_1 = 10\ \Omega$; $R_2 = 10\ \Omega$; $R_3 = 20\ \Omega$, $V = 240\ V$. *What current will be supplied?*

$R = R_1 + R_2 + R_3 = 10 + 10 + 20\ \Omega = 40\ \Omega$
$I = V/R = 240/40\ A = 6\ A.$

A source supplies 90 V to a series combination of 20 Ω and 60 Ω. What extra series resistance would limit the current to 1 A? [169]

Note: In all but the simplest problems, it will help you to sketch the circuit.

Tests on parallel circuits show these points.

In a parallel circuit
 (i) The potential difference across each element is the same.
 (ii) The current entering any point equals the current leaving it.
(iii) The sum of the currents in the elements is the current supplied.

For Fig. 14.13(b), these give

(a) $\qquad V = V_1 = V_2 = V_3$

(b) $\qquad I = I_1 + I_2 + I_3$

But $I = V/R$,

so $\qquad V/R = V_1/R_1 + V_2/R_2 + V_3/R_3$

This gives $\qquad V/R = V/R_1 + V/R_2 + V/R_3$

$1/R = 1/R_1 + 1/R_2 + 1/R_3$ for three elements in parallel

The reciprocal of the resistance of a set of elements in parallel is the sum of the reciprocals of their resistances.

Use these data with Fig. 14.13(b). $R_1 = 10\ \Omega$; $R_2 = 10\ \Omega$; $R_3 = 20\ \Omega$; $V = 240\ V$. *What current will be supplied?*

$1/R = 1/R_1 + 1/R_2 + 1/R_3 = 1/10 + 1/10 + 1/20\ \Omega^{-1} = 5/20\ \Omega^{-1}$
$\rightarrow R = 4\ \Omega$

$I = V/R = 240/4\ A = 60\ A.$

A source supplies 90 V to 20 Ω and 60 Ω in parallel. What extra series resistance is needed to limit the current to 1 A? [256]

What will be the value of I *in the circuit of Fig. 14.14?*

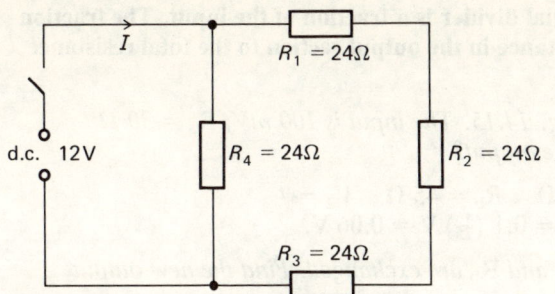

Fig. 14.14

$R_1 = 24\ \Omega \quad R_2 = 24\ \Omega \quad R_3 = 24\ \Omega \quad R_{123} = ?$
$R_{123} = R_1 + R_2 + R_3 = 72\ \Omega \quad R_4 = 24\ \Omega\ R = ?$
$1/R = 1/R_{123} + 1/R_4 = 1/72 + 1/24\ \Omega^{-1} \to R = 18\ \Omega$
$R = 18\ \Omega \quad V = 12\ V \quad I = ?$
$I = V/R = 12/18\ A = 2/3A.$

14.5 The potential divider

This is a most useful type of parallel circuit. In simple form we see it in Fig. 14.15.

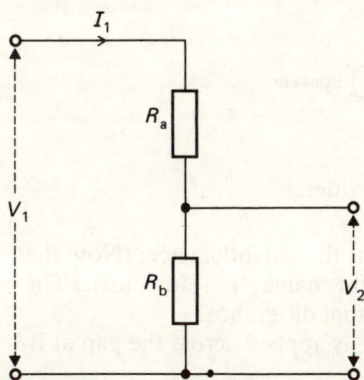

Fig. 14.15 The potential divider

The system is supplied with potential difference V_1. (This is its 'input'.) R_a and R_b form a series chain; the source will therefore supply a current I_1 given by $V_1/(R_a + R_b)$. The 'output' is V_2, the potential difference across R_b; it is given by $I_1 R_b$.

Thus $\qquad\qquad\qquad\qquad V_2 = I_1 R_b$

or $\qquad\qquad\qquad\qquad V_1 R_b/V_2 = R_a + R_b$

We can rewrite this as $\qquad\qquad V_2 = V_1\ (R_b/(R_a + R_b))$

The output of a potential divider is a fraction of the input. The fraction is the ratio of the resistance in the output section to the total resistance of the chain.

Use these data with Fig. 14.15. The input is 100 mV; $R_a = 30 \ \Omega$; $R_b = 45 \ \Omega$. What is the output?

$V_1 = 0.1 \ \text{V} \quad R_a = 30 \ \Omega \quad R_b = 45 \ \Omega \quad V_2 = ?$
$V_2 = V_1 R_b/(R_a + R_b) = 0.1 \ (\tfrac{45}{75}) \ \text{V} = 0.06 \ \text{V}.$

In the same circuit, R_a and R_b are exchanged. Find the new output.
[306]

To power the display of a watch, 15 V is required. The watch is supplied with a 30 V source. What potential divider would give the required output? [165]

A potential divider using a variable resistor gives an output p.d. that can be varied. Figure 14.16. shows one use – the volume control of a speaker.

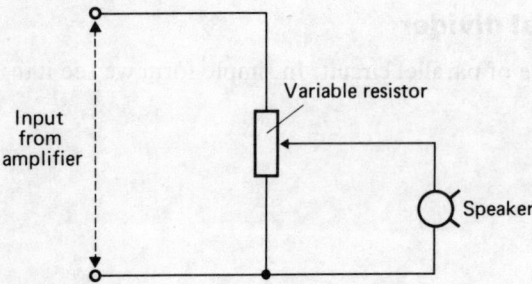

Fig. 14.16 Using a variable potential divider

Another variable potential divider is the **potentiometer**. (Note that some people call a variable resistor by this name, 'pot' for short.) The potentiometer is used to measure potential difference.

An unknown potential difference V is applied across the gap at B in Fig. 14.17.

As long as V is less than the 'driver' p.d., a 'balance' position of contact C can be found. Contact C is moved along the wire AD; when the meter reads zero, 'balance' is achieved.

At balance, the circuit compares to the potential divider in Fig. 14.18.

The unknown, V, equals and opposes the divider output.

Therefore $\qquad V = V_1 R_{\text{AC}}/(R_{\text{AC}} + R_{\text{CD}})$

or $\qquad\qquad V = V_1 (R_{\text{AC}}/R_{\text{AD}})$

The potentiometer wire is uniform (Fig. 14.17). Thus the

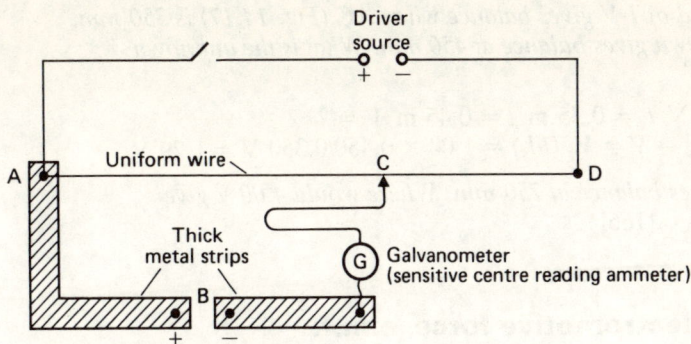

Fig. 14.17 The potentiometer

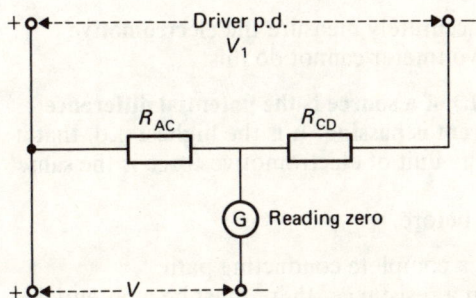

Fig. 14.18 The potentiometer as a potential divider

resistance of any section is proportional to the section length. We can therefore rewrite the above:

$V = V_1 \text{(AC/AD)}$

In most cases AD is 1.00 m; if we call AC l:

$V = V_1 l$

Thus we could measure any unknown V if the value of V_1 – the potential difference between A and D – is known. However, the value of V_1 cannot be known exactly because of the resistance of the wires, etc. It is therefore usual to apply first a standard, V_s, to B, and then to apply the unknown V. The two in turn give balance at lengths l_s and l. Then

---

$V/V_s = l/l_s$ (as $V = V_1 l$ and $V_s = V_1 l_s$).

Thus the value of V_1 (the 'driver' p.d.) no longer matters. (It must, however, be larger than either V or V_s.)

*A standard of 1 V gives balance when AC (Fig. 14.17) is 350 mm.
An unknown gives balance at 450 mm. What is the unknown
value?*

$V_s = 1.00$ V $l_s = 0.35$ m $l = 0.45$ m $V = ?$
$V/V_s = l/l_s \rightarrow V = V_s (l/l_s) = 1.00 \times 0.450/0.350$ V $= 1.29$ V

*9.00 V gives balance at 750 mm. Where would 4.00 V give
balance?*   [166]

14.6 Electromotive force, e.m.f.

Used with care, the potentiometer can measure an unknown potential
difference with great accuracy. All the same, it is clearly not as easy to
use as a voltmeter.

Its strength is that it can accurately measure the **electromotive
force** (e.m.f.) of a source. A voltmeter cannot do this.

**The electromotive force (e.m.f.) of a source is the potential difference
between its ends when no current is passing**. It is the highest p.d. that it
can supply, as we shall see. The unit of electromotive force is the same
as that of p.d. – the volt, V.

Recall some points made before.

(i) A direct current circuit is a complete conducting path.
(ii) When there is a current in a resistance, there must be a potential
difference between its ends.

Figure 14.19 completes Fig. 14.6; it shows details of the source –
any source. This itself has resistance, r, as well as e.m.f., E.

When any source is used, a current I passes in it as well as in
the rest of the circuit. So part of the source's e.m.f. must supply a
potential difference, v, inside it. This p.d., often called the potential
drop, is used to drive the current I through the source resistance r.

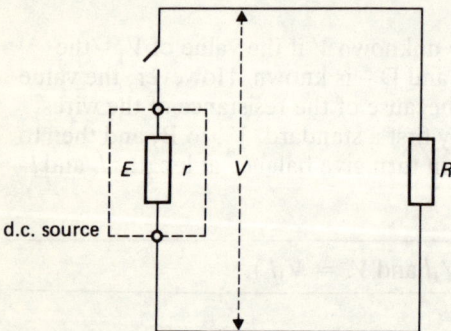

Fig. 14.19 A complete d.c. circuit

$$v = Ir.$$

For the rest of the circuit $V = IR$

So $V + v = I(R + r)$

But $(V + v)$ is the electromotive force of the source – the total p.d. supplied.

So

$$E = I(R + r)$$

The electromotive force of a direct current source is the product of the current supplied and the total circuit resistance.

A source of e.m.f. 12 V and resistance 2 Ω supplies a 10 Ω circuit. What is the current?

$E = 12$ V $r = 2$ Ω $R = 10$ Ω $I = ?$
$E = I(R + r) \rightarrow I = E/(R + r) = 12/(10 + 2)$ A $= 1$ A

The same source now supplies two 10 Ω resistors in parallel. What is the current?   [292]

How is electromotive force measured? The potentiometer provides an accurate and fairly easy method. We connect a standard e.m.f. E_s in gap B (Fig. 14.17) and find the balance length l_s of AC. We now replace the standard with the unknown e.m.f. E; the new balance length is l.

Then

$$E/E_s = l/l_s$$

A standard source of e.m.f. 1.10 V gives balance on a potentiometer when AC in Fig. 14.17 is 550 mm. Balance with an unknown is at 505 mm. Find the unknown e.m.f.

$E_s = 1.10$ V $E = ?$ $l_s = 0.550$ m $l = 0.505$ m
$E/E_s = l/l_s \rightarrow E = E_s (l/l_s) = 1.10 \times 0.505/0.550$ V $= 1.01$ V

A standard e.m.f. of 0.85 V gives balance on a potentiometer when AC in Fig. 14.17 is 170 mm. A second source, e.m.f. 3.3 V, is now used. What will be the balance length?   [344]

The potentiometer is of value when one wishes to measure e.m.f. because it takes no current from the source under test – the meter reads zero at balance. Recall that electromotive force is the potential difference between the source terminals when no current is taken.

Figure 14.20 outlines a second method for finding the e.m.f. of a source. It takes longer and does not give such a good result. On the other hand it uses standard equipment.

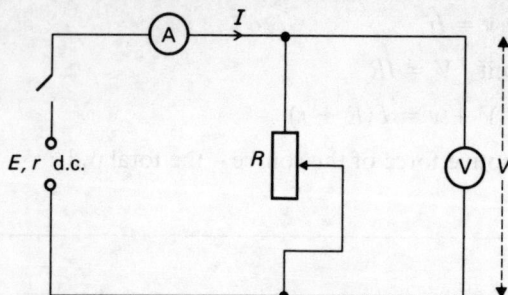

Fig. 14.20 To find source e.m.f. using ammeter and voltmeter

A number of readings of I and V are taken as the value of R is changed. These readings are plotted on a potential difference – current graph. It will look like that in Fig. 14.21.

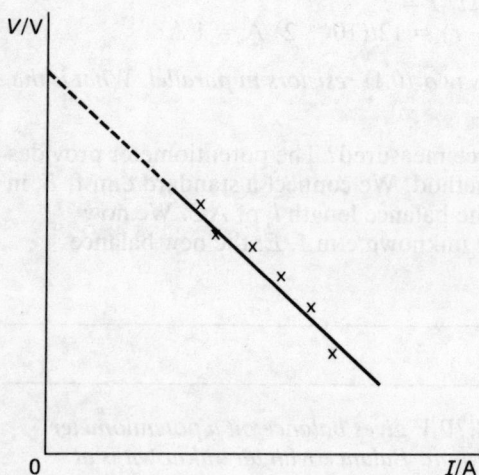

Fig. 14.21

When the graph is produced back to the V-axis, we find the potential difference when $I = 0$ – this is the source e.m.f. (We can also use the slope to give source resistance.)

As the meters are not perfect, this method cannot give a really correct answer.

14.7 Some more questions

1. *Copy and complete Table 14.1. It refers to a number of circuit elements.*

Table 14.1

	Between ends	Resistance	Current passed	
(a)	10 V	100 Ω		[142]
(b)	5 kV	250 Ω		[157]
(c)	250 V		5 A	[352]
(d)	100 mV		5 A	[216]
(e)		10 Ω	2 A	[192]
(f)		30MΩ	90 mA	[167]

2. *A certain copper sample has resistance 10 Ω. Find the resistance of a second copper sample of twice the length and twice the radius.* [246]
3. *A room has two ceiling lamps, connected in parallel. There is a single switch. Give a drawing of the system in practice and a circuit diagram.* [345]
4. *Describe the circuit of Fig. 14.22 in words, so that someone else could build it without a diagram.* [416]

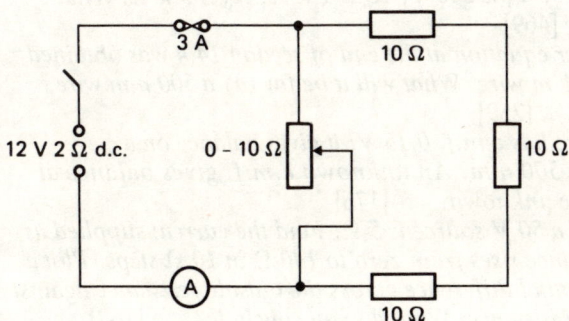

Fig. 14.22

5. *What will the meter in Fig. 14.22 read if the variable resistor is set at 7.5 Ω?* [174]
6. *In Table 14.2 are potential difference and current data for a certain circuit element. Plot a graph; compare it with Fig. 14.10. What type of element do you think this is?* [488]
7. *5 Ω and 10 Ω are joined in series; the result is joined in parallel to 15 Ω across a 90 V p.d. What is the current?* [502]
8. *5 Ω and 10 Ω are joined in parallel; the result is joined in series with 15 Ω across a 90 V p.d. What is the current?* [223]
9. *Find the resistance of each source.*
   (a) *A 12 V 'battery' which supplies 10 A to a 0.2 Ω load.* [402]

(b) *A 250 V generator feeding 25 A to a 5 Ω load.* [319]
(c) *A 1.5 V photocell from which a 50 Ω load takes 15 m A.* [145]

Table 14.2

V/V	I/mA
0	1.0
10	2.0
20	4.0
30	7.0
40	10.0
50	13.0
60	16.0
70	17.5
80	17.9
90	18.0

10. *Draw the eight ways in which 1 Ω, 2 Ω and 3 Ω can be joined. Find the total resistance in each case. List the answers in order.* [265]
11. *Refer to Fig. 14.15.* V_1 *is 240 V;* R_a *is 1 k Ω;* R_b *is 3 k Ω. What is the output?* [469]
12. *The potentiometer equation at the end of section 14.4 was obtained for the case of a 1 m wire. What will it be for (a) a 500 mm wire, (b) a 2 m wire?* [259]
13. *A standard source has e.m.f. 0.15 V; it gives balance on a potentiometer at 500 mm. An unknown e.m.f. gives balance at 350 mm. Find the unknown.* [176]
14. *The resistance of a 50 V source is 5 Ω. Find the current supplied as the outside resistance rises from zero to 140 Ω in 10 Ω steps. Plot a graph of the potential difference across the outside resistance against the value of that resistance. What do you conclude?* [168]
15. *Table 14.3 gives data from an experiment like that described in Fig. 14.20. Plot a p.d. – current graph; from it find the e.m.f. and resistance of the source.* [277]

Table 14.3

Reading	1	2	3	4	5	6
I/A	1.0	1.1	1.2	1.5	2.0	3.0
V/V	10.0	9.8	9.6	9.0	8.0	6.0

16. *What was the value of* R *(Fig. 14.20) for each reading given in the last question?* [433]
17. *Describe, with a diagram, how you would attempt to find the resistance of saturated brine in a beaker. On what factors do you think your answer would depend?* [492]

14.8 Objectives

Note: The first nine objectives are from Level 1.
When you have studied this chapter, you should be able to

(1) state that like charges repel each other and unlike charges attract each other;
(2) describe potential difference, current and resistance in terms of how charges behave in matter;
(3) state Ohm's law and recognise cases of non-ohmic behaviour;
(4) describe the measurement of potential difference, current and resistance using special meters, a multimeter and an oscilloscope (for p.d.);
(5) state the units of potential difference, current and resistance and solve simple problems relating these measures;
(6) state that sample resistance depends on material, length, cross-section area and temperature;
(7) use preferred symbols for elements in circuit diagrams;
(8) define series and parallel combinations of elements;
(9) derive and use expressions giving the resistance of these;
(10) explain the potential divider circuit and its forms and solve simple problems on these;
(11) describe the use of the potentiometer to compare p.d.'s and e.m.f.'s;
(12) define e.m.f. as p.d. when no current is supplied;
(13) relate e.m.f., source and circuit resistances, and current supplied, and solve simple problems on these;
(14) sketch a typical graph of the p.d. across a source against the current supplied and use it to estimate e.m.f.

Practical: You should also have practice in building circuits from diagrams, and in using the techniques discussed.

Chapter 15

The motor effect

15.1 Magnetism and electricity

Magnetism and electricity relate closely to each other. We have
already seen that both are concerned with light and other waves of the
same kind – electromagnetic radiation. In this chapter and the next two
we shall meet many other examples of this close relationship between
electricity and magnetism.

Figures 15.1(a) and (b) show experiments which relate the two. In
one, **moving magnetism causes an electric effect**; in the other **moving
electricity causes a magnetic effect.**

In this chapter we look at the '**motor effect**' of electricity.

When electric charge crosses a magnetic field, it experiences a force.

The effect has many uses.

Before we look at some of these, we need some basic ideas about
magnetism.

(i) Tests show that two **types of magnetic pole** exist. They are called
n-pole and s-pole. They always appear together in pairs.

(ii) **Like poles repel each other; unlike poles attract each other.**

(iii) The forces between poles can act without contact.

(iv) We say that a **magnetic force field** surrounds each pole. **A magnetic
force field is a region of space in which magnetic forces can be
observed**. We detect magnetic force fields with a **compass** (as in
Fig. 15.1(b)); we can also see them using **iron dust**.

(a) Moving magnetism causes an electric effect

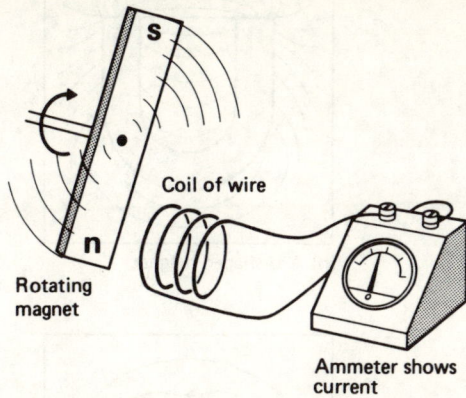

Coil of wire

Rotating magnet

Ammeter shows current

(b) Moving electricity causes a magnetic effect

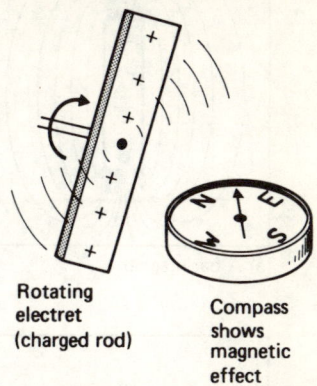

Rotating electret (charged rod)

Compass shows magnetic effect

Fig. 15.1

Iron dust forming field pattern

Glass sheet

Compass

n   s

Magnet

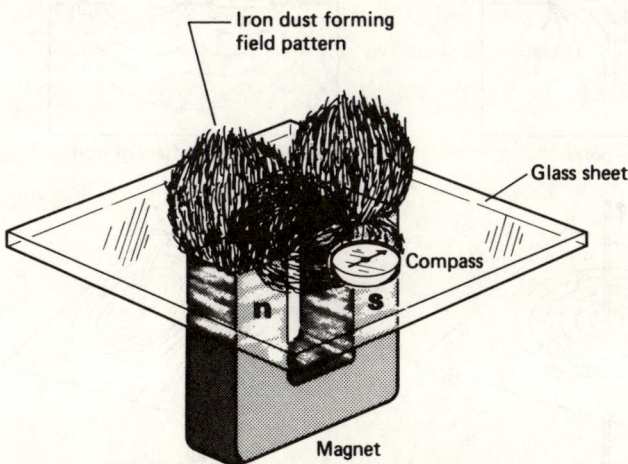

Fig. 15.2 The field of a U-shaped magnet

Figure 15.2 shows the three-dimensional field of a strong U-shaped steel magnet. The picture illustrates some of these points. Here we have the two types of pole, an n-pole and an s-pole. The force of attraction between them is shown by the way the iron dust links them. Each bit of iron dust is affected by the magnetic force of the poles, although it does not touch them.

Cross-sections of various magnetic force fields are sketched in Fig. 15.3. We draw lines to show fields as there is no other easy way – but fields do not consist of lines; they fill the space concerned. The lines show (by their direction) the direction of the field, and (by their

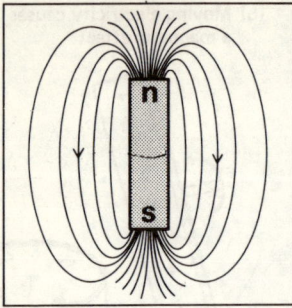

(a) A bar magnet

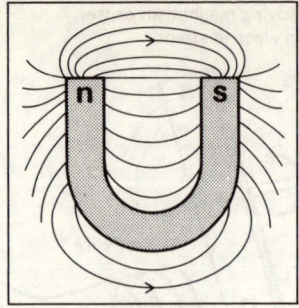

(b) A U-shaped magnet

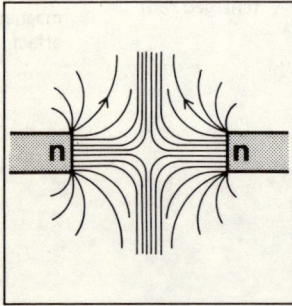

(c) Two close like poles

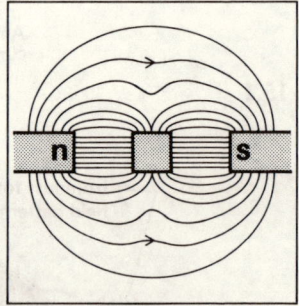

(d) The focusing effect of iron

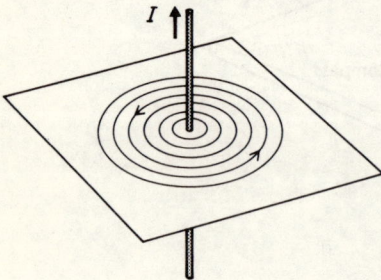

(e) An electric current

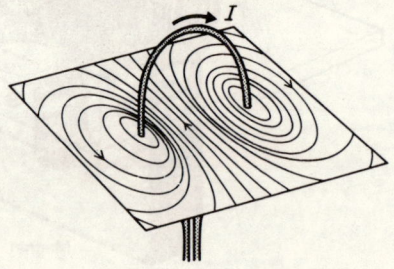

(f) The current in a loop

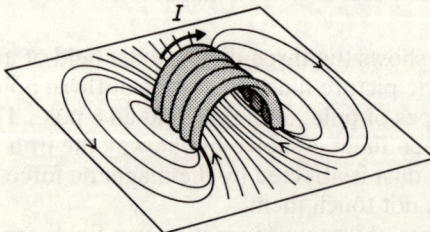

(g) The current in a coil

Fig. 15.3 Some magnetic fields

closeness) the field strength. This is because magnetic field strength is a vector, like other force field strengths.

Magnetic fields are found in two types of situation. Permanent magnets (as in Fig. 15.3(a) to (d)) do not easily lose their fields; they are used in speakers, toys, meters, refrigerator doors, dynamos and many other things. Electromagnetic fields (like those in Fig. 15.3(e), (f) and (g)) exist only when charge flows – they can be switched on and off.

15.2 The motor effect

When electric charge crosses a magnetic field, it experiences a force.

On what factors would you expect the force to depend?

The force depends on the magnetic field strength ('flux density') (B). In fact the two are proportional – if the field strength is doubled, so is the force; if there is no field, there is no force.

$F \propto B$.

The force depends on the charge (Q). Again the two are proportional – if the charge doubles, so does the force; if there is no charge, there is no force.

$F \propto Q$.

The force depends on the velocity with which the charge crosses the field. Here too the measures are proportional – if the velocity doubles, so does the force; if the object is not moving, there is no force.

$F \propto v$.

(Note that we speak of velocity rather than speed. In this book we restrict ourselves to cases where v is at 90 ° to B. In fact, the angle between v and B also affects the force – if this is zero, so is the force.)

The above three proportionalities combine like this. (The constant of proportionality is one.)

$F = BQv$

The force on a charge crossing a magnetic field is the product of the field strength, the charge and the velocity.

(Note that magnetic field strength is also called **flux density**. 'Flux' is an old name for 'lines of force' so flux density means the number of lines of force per unit section area of the field).

Force is a vector (section 1.3). In which direction does it act? As Fig. 15.4 shows, the force is at 90 ° to both B and v.

Figure 15.4(a) gives the three vectors. Figure 15.4(b) shows the

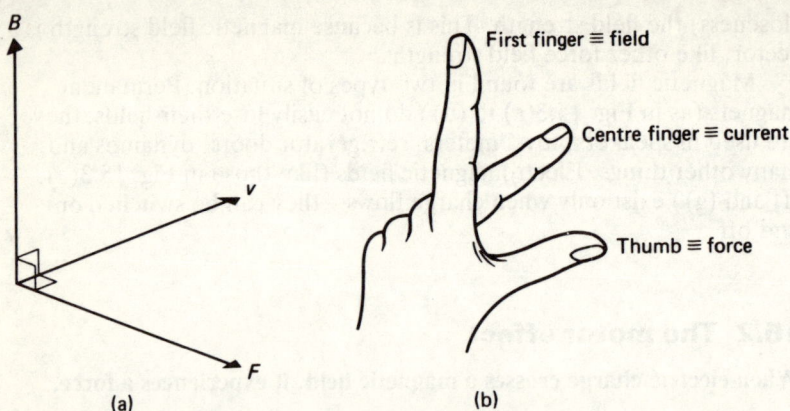

First finger ≡ field

Centre finger ≡ current

Thumb ≡ force

(a)   (b)

Fig. 15.4 The directions of field, velocity and force

left-hand motor rule; we can always use this to tell us the direction of the force.

Be clear that Fig. 15.4(a) shows a three-dimensional relationship between vectors. Here we have the three vectors – B, magnetic field strength, v, the velocity of the charge, and F, the motor force on the charge – all at 90 ° to each other. In other words if the charge is moving north through a vertical magnetic field, an east–west force will act on it. These three directions are shown by the left hand in Fig. 15.4(b).

We can state the left-hand motor rule like this. **The thumb, first, and second fingers of the left hand point at 90° to each other; if the first finger points along the field, and the centre finger points the way the (positive) charge is moving, the thumb shows the way the force acts.**

In brief:

First finger ≡ field   Centre finger ≡ current   Thumb ≡ force

Note that this applies to a positive charge; a negative charge will be forced the other way. (A negative charge coming south is equivalent to a positive charge moving north.)

To use the equation $F = BQv$ in given cases, we need to know the units of the measures concerned.

F is the force on the charge; the unit is the newton, N.

B is the field strength; the unit is the tesla, T. (See section 15.3.)

Q is the charge; its unit is the coulomb, C. (That is the charge passed by one ampere in one second.)

v is the charge velocity; the unit is the metre per second, m s^{-1}.

Note that the coulomb is large; it equals the charge carried by about 6×10^{18} electrons. (The charge of one electron is about 1.6×10^{-19} C.)

An electron travels north at 100 000 km s^{-1}. It enters a 10 T field directed east. (a) Which way will the charge be forced? (b) How large is

the force? (c) *What acceleration results?* The electron charge is 1.6×10^{-19} C; its mass is 9.1×10^{-31} kg.

(a) Use the left-hand rule. The electron is forced upwards.
(b) $B = 10$ T $\quad Q = 1.6 \times 10^{-19}$ C $\quad v = 10^8$ m s^{-1} $\quad F = ?$
$\quad F = BQv = 10 \times 1.6 \times 10^{-19} \times 10^8$ N $= 1.6 \times 10^{-10}$ N
(c) $m = 9.1 \times 10^{-31}$ kg $\quad F = 1.6 \times 10^{-10}$ N $\quad a = ?$
$\quad a = F/m = 1.6 \times 10^{-10}/9.1 \times 10^{-31}$ m s$^{-2} = 1.8 \times 10^{20}$ m s^{-2}

At what speed should a 100 μC charge cross a 100 T field to experience a 10 N force? [263]

15.3 The force on a conductor

The 'motor effect' on free moving charges has a number of uses. In TV tubes, for instance, the electron beam is moved to reach all parts of the screen by changing the electromagnetic fields in the coils. Figure 15.5 shows this.

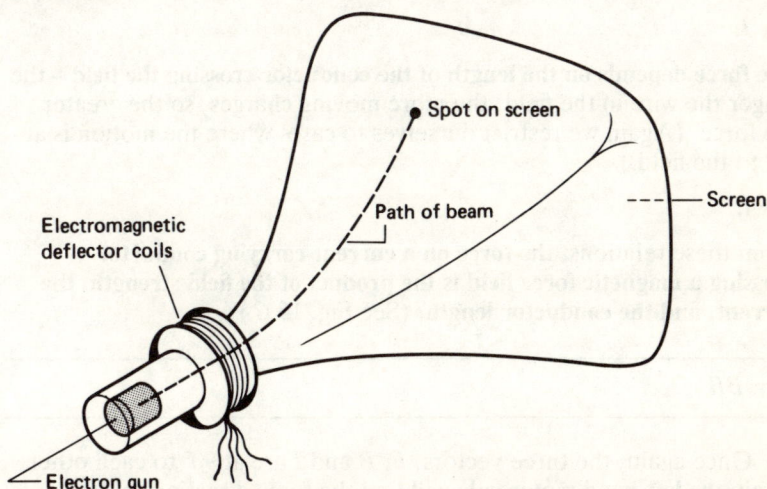

Fig. 15.5 Electromagnetic deflection in a TV tube

The Earth's magnetic field is not strong – but it is vast. It can deflect the dangerous charged particles from the Sun (the 'solar wind') towards the poles. (There they cause the **aurorae**.)

Most uses of the motor effect, however, apply to charges moving along a conductor. In other words, they involve currents in wires. Figure 15.6 shows this for a *positive* charge flow. The charges are forced upwards; they cannot leave the wire – so the wire is forced upwards.

The force on each charge is BQv (section 15.2). The force F on the wire also depends on three factors.

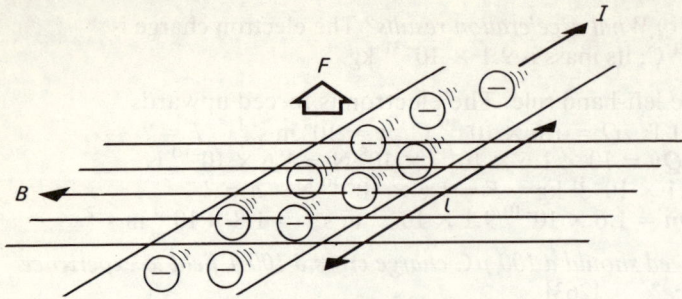

Fig. 15.6 Motor force on a current-carrying wire

The force depends on the magnetic force field strength (flux density) – just as before:

$F \propto B$.

The force depends on the current – this relates to the number and size of the charges passing, and their mean velocity.

$F \propto I$.

The force depends on the length of the conductor crossing the field – the longer the wire in the field, the more moving charges, so the greater the force. (Again we restrict ourselves to cases where the motion is at 90° to the field.)

$F \propto l$.

From these relations, **the force on a current-carrying conductor crossing a magnetic force field is the product of the field strength, the current, and the conductor length.** (See Fig. 15.6.)

$$F = BIl$$

Once again, the three vectors, *F, B* and *I* are at 90° to each other. Again **the left-hand motor rule** will find the force direction for a *positive* charge flow. (See Fig. 15.4, replacing the charge motion with current.)

We can use $F = BIl$ to define magnetic field strength.

$$B = F/Il$$

The strength of a magnetic force field is the force it applies to a conductor of unit length crossing it and carrying unit current.

From this, the unit of B is the newton per ampere metre $(\mathrm{NA^{-1}\,m^{-1}})$; we call this the **tesla**, T.

One tesla is the strength of the field able to apply one newton to a one metre conductor crossing it and carrying one ampere.

A circuit carries 5 A. 100 mm of the wire crosses a 5 T field. Find the force on the wire.

$I = 5\,A \quad l = 0.1\,m \quad B = 5\,T \quad F = ?$
$F = BIl = 5 \times 5 \times 0.1\,N = 2.5\,N$

Figure 15.7 shows a good demonstration of the effect. In one experiment I was 2.5 A and l was 50 mm. The roller, of mass 25 g, accelerated at 100 mm s^{-2}. Find the field strength. Also copy the sketch, showing the way the roller moved. [441] [175]

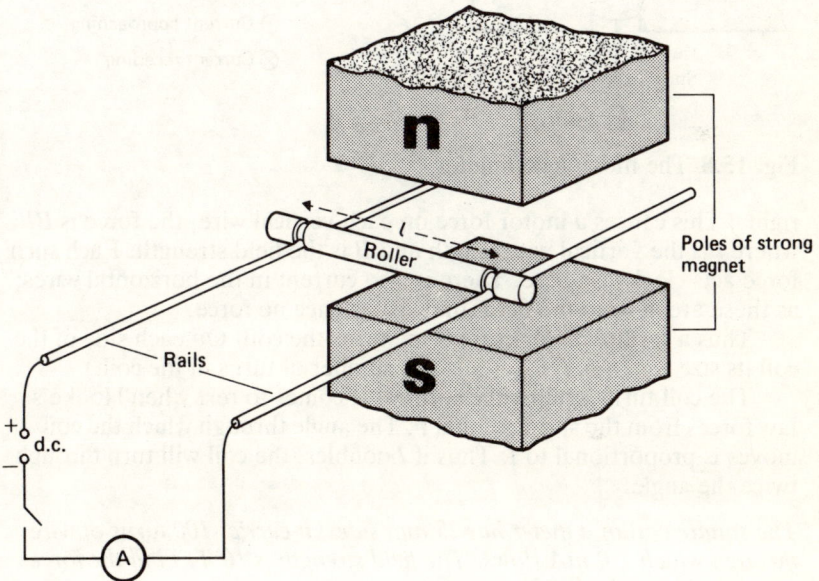

Fig. 15.7

15.4 The moving coil meter

This design is widely used for electricity meters. You may have worked with moving coil meters while going through the last chapter.

The current concerned passes through the coil in Fig. 15.8(a). This is wound on a plastics 'former' (= base) of square section; it hangs in the gap between the strong magnet poles and the iron core. The magnet and core produce a radial field. This is shown dotted in Fig. 15.8(b). (Compare this with Fig. 15.3(d).) Thus, however, the coil turns, its vertical sides carry current at 90° to the field.

When there is a current *I* through the coil, it passes along each vertical wire. (In our drawings, it goes up on the left and down on the

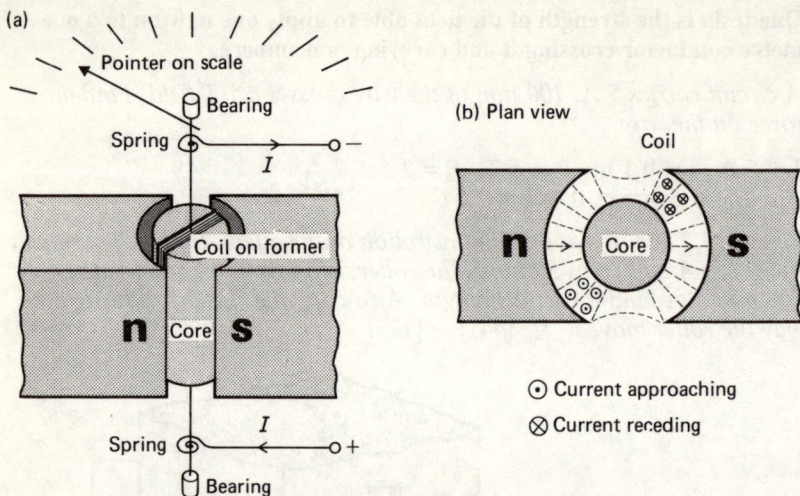

Fig. 15.8 The moving coil meter

right.) This causes a motor force on each vertical wire; the force is BIl, where l is the vertical wire length and B is the field strength. Each such force acts clockwise here. There is also current in the horizontal wires; as these are at $0°$ to the field, they experience no force.

Thus a resultant clockwise force turns the coil. On each side of the coil its size is $F = NBIl$. (N is the number of turns in the coil.)

The coil turns against the springs. It comes to rest when Hooke's law forces from the springs equal F. The angle through which the coil moves is proportional to I. Thus if I doubles, the coil will turn through twice the angle.

The square coil of a meter has 25 mm sides; it carries 100 turns of wire through which 1.0 mA flows. The field strength is 10 T. Find the force on each side of the coil.

$l = 0.025$ m $N = 100$ $I = 10^{-3}$ A $B = 10$ T $F = ?$

$F = NBI\,l = 100 \times 10 \times 10^{-3} \times 0.025$ N
$= 0.025$ N $= 25$ mN

15.5 The electric motor

This is the most obvious use of the electric motor effect! The motor does not differ too much from the meter (section 15.4). This time, however, the coil's turning is not opposed by a spring – the coil will turn as long as there is a force on it.

Figure 15.9 outlines the arrangement in a small motor. As with the meter, there is a field produced by the poles of a strong magnet; again

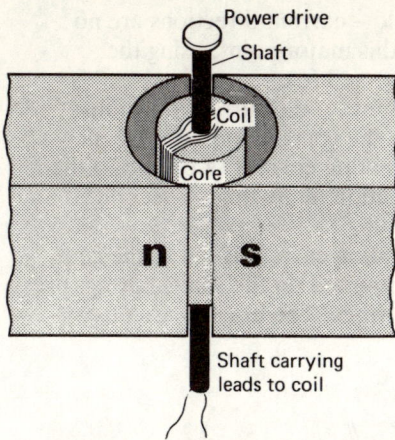

Fig. 15.9 Inside a small electric motor

this is made radial by an iron core.

As before, when there is a current in the coil, it will turn. Now it is attached to the core and shaft; these will therefore turn too. The turning forces will move the coil from (a) in Fig. 15.10, through (b) and (c). However, as soon as it passes (c) the forces will be in the

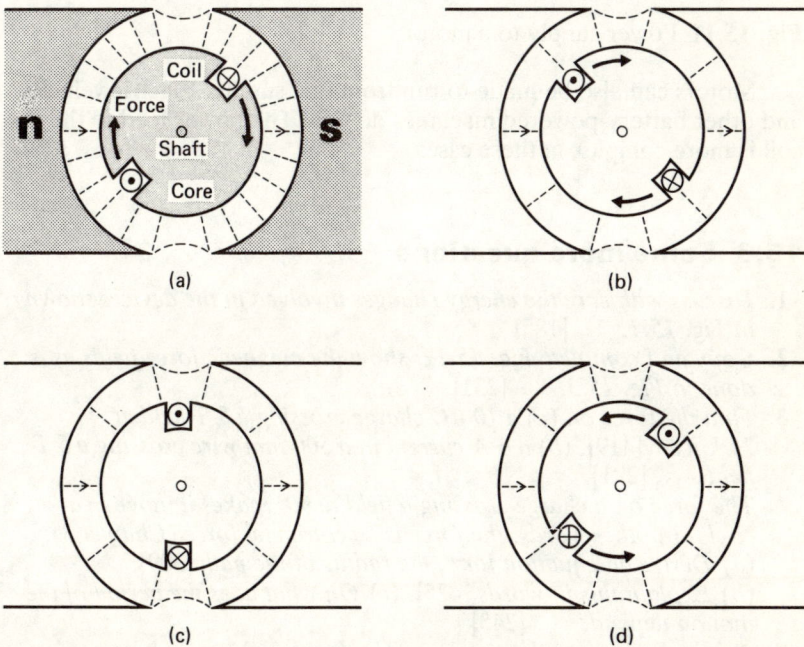

Fig. 15.10

reverse direction. This is because field – current directions are no longer the same. You should check this major point, using the left-hand motor rule, on Fig. 15.10(b) and (d).

To keep the coil turning the same way, the current flow must reverse when the coil passes through the (c) position. There is no problem here – mains power is **alternating current** (a.c.) rather than direct current (d.c.) – it changes direction many times a second. It follows a sine wave pattern.

To make the motor work, alternating current is fed to the coil through two 'slip rings'. These are mounted on the shaft. Springs press sliding contacts ('brushes') on to the rings. The brushes are often carbon blocks. (See Fig. 15.11.)

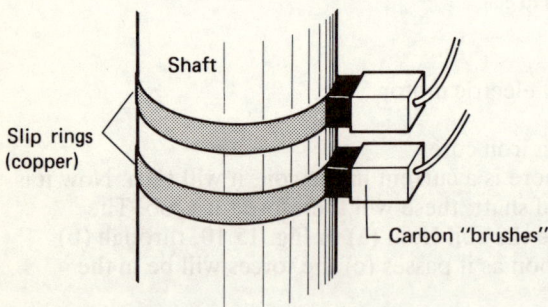

Fig. 15.11 Power supply to a motor

Motors can also be made to run from d.c. supply. Electric vehicles and other battery-powered machines do this. The power feed to the coil is more complex in these cases.

15.6 Some more questions

1. *Discuss with care the energy changes involved in the devices shown in Fig. 15.1.* [185]
2. *Copy and complete Fig. 15.12; show the magnetic force fields as is done in Fig. 15.3.* [231]
3. *Find the force on (a) a 10 μC charge crossing a 5 T field at 15 km s^{-1} [419], (b) a 6 A current in a 500 mm wire crossing a 5 T field.* [291]
4. *The force on a charge crossing a field at 90° makes it move in a circle. In other words, the force is a centripetal force (Chapter 3). (a) Derive an equation for r, the radius of the path [309]. (b) Explain this in words [425]. (c) On what does the period of the motion depend?* [245]
5. *An electron leaves the gun in a TV tube at 20 Mms^{-1}. It passes for 500 mm across the Earth's field (20 μT) before hitting the screen.*

How far is the spot on the screen from where it would be if there were no field? The electron charge is 1.6×10^{-19} C; its mass is 9.1×10^{-31} kg. [353]

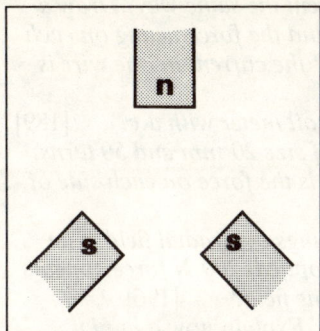

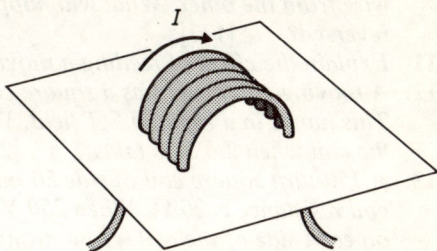

(a) Three poles   (b) The current in a coil

Fig. 15.12

6. *Because of a storm during launch a space-craft carries a 50 C charge. What is the force on it as it orbits the Earth at 8 km s^{-1} through the Earth's field (10 μT)?* [453]
7. *Figure 15.13 shows a vertical section of a pipe carrying liquid metal. A strong field is applied at 90° to the picture while d.c. passes between the plates. The result is a pumping effect. Explain how this pump works and why it is often used in such cases.* [428]
8. *There is a 2 T field between the poles of a magnet. What is the force on a 15 mm length of wire carrying 3 A, when the wire is (a) at a right angle to the field [273], (b) parallel to the field?* [483]

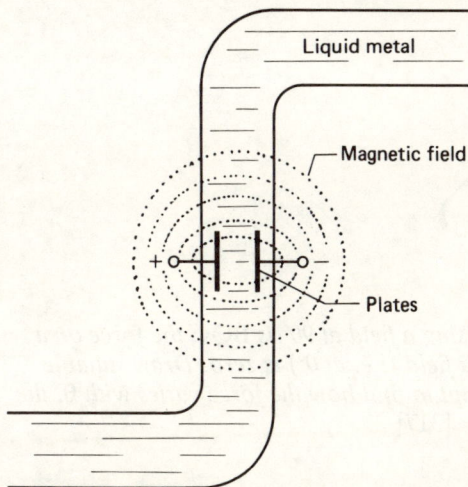

Liquid metal

Magnetic field

Plates

Fig. 15.13

9. *A 100 mm wire carrying 10 A lies across a magnetic field. A 30 g load on the wire stops it moving upwards. How strong is the field?*   [257]

10. *Two close parallel wires each carry current the same way. Draw a force field picture from which you can find the force acting on each wire from the other. What will happen if the current in one wire is reversed?*   [177]

11. *Explain the effect of feeding a moving-coil meter with a.c.*   [189]

12. *A moving coil meter has a square coil of size 20 mm and 50 turns. This hangs in a radial 0.5 T field. What is the force on each side of the coil when 2.5 A is taken?*   [232]

13. *A 150-turn square coil of side 50 mm hangs in a radial field. The coil resistance is 20 Ω. When 250 V is applied, 200 N force appears on each side of the coil. How strong is the field?*   [196]

14. *Figure 15.14 shows a speaker in section. Explain how a hum is produced when the speaker is supplied with a.c.*   [282]

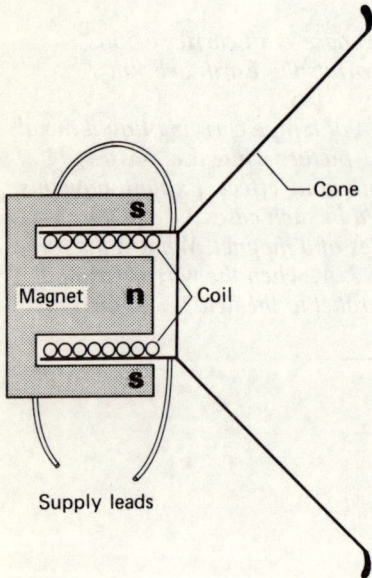

Fig. 15.14

15. *The force on a charge crossing a field at 90° is BQv; the force on a charge moving parallel to a field (i.e. at 0°) is zero. Draw suitable sketches; use them to attempt to find how the force varies with θ, the angle between B and v.*   [317]

15.7 Objectives

Note: The first objective is from Level 1.
When you have studied this chapter, you should be able to

(1) state the basic concepts of magnetism;
(2) sketch simple magnetic and electromagnetic force fields;
(3) state that the force on a moving charge in a field depends on charge, field strength, velocity and angle;
(4) explain and use the relation $F = BQv$;
(5) state that the force on a current-carrying conductor in a field depends on current, field strength, length and angle;
(6) explain and use the relation $F = BIl$;
(7) from this define B and its unit, the tesla;
(8) sketch, label and describe the movement of a moving-coil meter;
(9) calculate the force on one side of a square coil at 90° to a field;
(10) describe the structure and action of a simple a.c. motor.

Chapter 16

Generating electric energy

16.1 Electromagnetic induction

There are many ways of providing an electromotive force (e.m.f.) – a source of electric energy. In each case some other energy form leads to electric effects.

For instance – a car **battery** and a dry **cell** provide electric energy from **chemical** action.

– A **thermocouple** makes temperature differences – differences in **thermal** energy content – show as electricity.

– A **photocell** absorbs **light** waves to make electricity.

– A **microphone** absorbs sound waves with the same effect.

Perhaps you can think of other cases.

Of those mentioned, only the first – chemical sources – can produce large currents. (The starting motor of a car may take 100 A or more from the battery.) They cannot do this for long, however.

To produce large currents *for a long time*, only **mechanical energy** sources are used at present. These **generators** are based on what is called **electromagnetic induction**.

Look again at Fig. 15.1.

When the meter in the circuit shows a reading there must be a current in it. This can happen only if there is an electromotive force (e.m.f.) in the circuit. The motion of the magnet by the coil must have produced this e.m.f. The appearance of an e.m.f. in such a situation is electromagnetic induction.

Figure 16.1 shows how we can test this effect. The best meter to use is a centre-reading galvanometer. (As described in Fig. 14.17, this is a sensitive ammeter.)

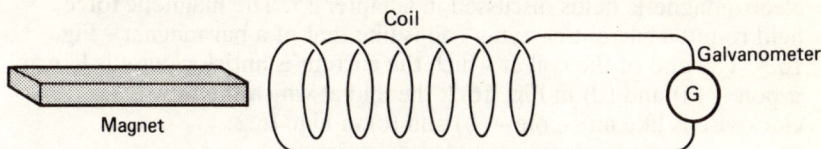

Fig. 16.1 Testing electromagnetic induction

Here is what tests show.

1. (a) When the magnet moves into or out of the coil, the meter records charge flow.
   (b) When the coil moves to or from the magnet, the meter records charge flow.
   (c) When neither coil nor magnet move, the meter reads zero.

When a conductor moves relative to a magnetic force field, an e.m.f. is induced.

2. (a) When the magnet moves slowly to or from the coil, the meter reading is small.
   (b) When the coil moves slowly to or from the magnet, the meter reading is small.
   (c) When the magnet moves quickly to or from the coil, the meter reading is large.
   (d) When the coil moves quickly to or from the magnet, the meter reading is large.

The size of the induced e.m.f. depends on the relative velocity between conductor and field.

3. See Fig. 16.2.
   (a) When the n-pole enters the coil, the current is anticlockwise round the coil.
   (b) When the s-pole enters the coil, the current is clockwise round the coil.

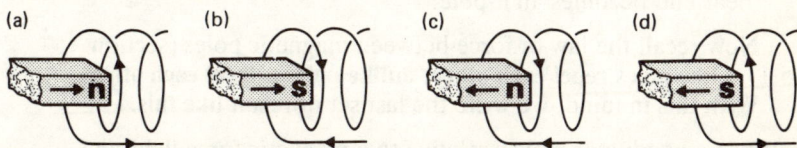

Fig. 16.2 The direction of induced e.m.f.

(c) When the n-pole leaves the coil, the current is clockwise round the coil.

(d) When the s-pole leaves the coil, the current is anticlockwise round the coil.

To see the meaning of this last set of tests, look again at the electromagnetic fields discussed in Chapter 15. The magnetic force field round a current-carrying coil is like that of a bar magnet – Fig. 16.3. The end of the coil at which the current is anticlockwise is like an n-pole – (a) and (d) in Fig. 16.2; the end at which the flow is clockwise is like an s-pole – (b) and (c) in Fig. 16.2.

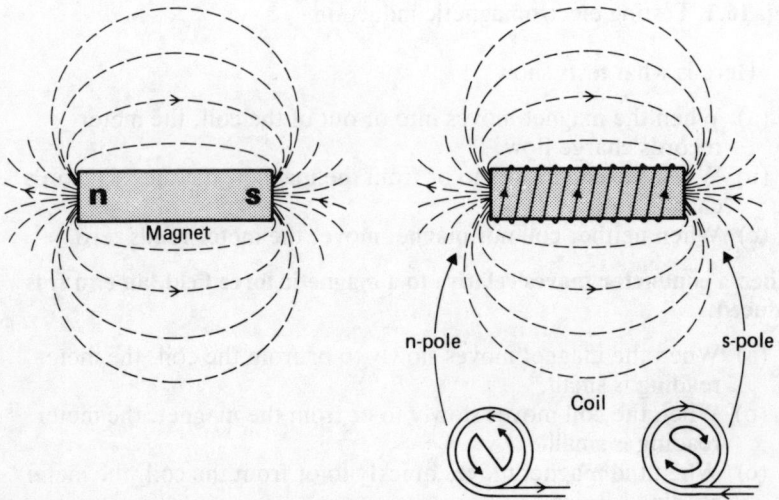

Fig. 16.3 The field of a current in a coil

So we can rephrase the last set of results:

(a) When the n-pole enters the coil, charge flows so that the coil's near end becomes an n-pole.

(b) When the s-pole enters the coil, charge flows so that the coil's near end becomes an s-pole.

(c) When the n-pole leaves the coil, charge flows so that the coil's near end becomes an s-pole.

(d) When the s-pole leaves the coil, charge flows so that the coil's near end becomes an n-pole.

Now recall the law of force between magnetic poles (section 15.1) – **like poles repel each other; unlike poles attract each other**. With this in mind, we write the last set of result like this.

When a conductor moves relative to a magnetic force field, the current induced sets up a field to oppose the motion.

In other words, **the direction of the induced e.m.f. is to oppose the relative motion between conductor and field.**

We would expect this. When charge moves through the circuit, setting up an e.m.f., energy is needed. We cannot create energy from nothing (section 4.5); our muscles must do more work (exert more force) to move the conductor in the field than if the field were not there. When we do the tests described here, we cannot feel this extra work – but recall how much harder it is to pedal a cycle when the dynamo is in use. The energy comes from that used to move the conductor in the field.

The three statements that arise from these tests are the **laws of electromagnetic induction.**

(i) When a conductor moves relative to a magnetic force field, an e.m.f. is induced.
(ii) The size of the induced e.m.f. depends on the relative velocity between conductor and field.
(iii) The direction of the induced e.m.f. is to oppose the relative motion between conductor and field.

The first two laws are due to Faraday; the third is **Lenz's law.** (Lenz is a German name; pronounce it 'Lents'.)

16.2 Mutual induction

In the last section we derived the laws of electromagnetic induction by a series of tests. These involved moving the force field of a magnet with respect to a conductor. (See Fig. 16.1.)

There is no reason for the source of the field to be a magnet. As we saw in Fig. 16.3, the field of a current in a coil is almost the same.

We could therefore repeat the experiments with an electromagnet – a d.c. in a coil (the 'primary') – rather than a permanent magnet. Just the same effects are obtained; Faraday's and Lenz's laws are still valid.

But we don't even need to move the primary! Instead we can simply switch the current on and off as desired. With the primary supplied as in Fig. 16.4(a), switching on is like bringing up an n-pole – there was no n-pole before, but there is now. Switching off is like taking the n-pole away. With the primary supplied as in Fig. 16.4(b), switching on is like bringing up an s-pole; switching off is like taking the s-pole away.

Again, we find the same effects in the other coil, the 'secondary'; again the laws of induction need no change. We can describe the effects like this.

A changing current through the primary coil causes a changing magnetic field in the secondary; this induces an e.m.f. in the secondary.

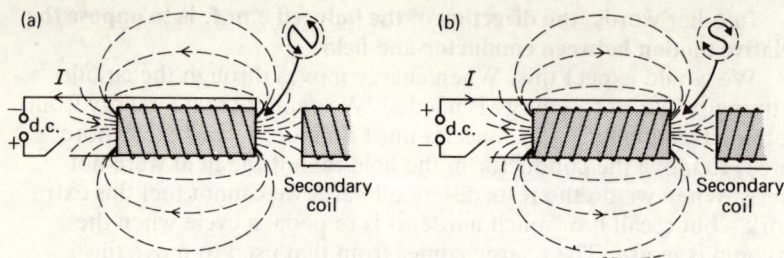

(a) (b)

I I

d.c. d.c.

Secondary coil Secondary coil

Fig. 16.4 Electromagnetic induction without motion

The reverse is true, too – we have the same effects if the secondary is powered, and the primary is joined to the meter.

The term flux linkage is often used to mean the number of lines 'linking' the circuit containing the coil. (I observed in Section 15.2 that flux was used to mean lines of force). In this case the induced e.m.f. can be seen to be proportional to the rate of change of flux linkage.

These effects, and others like them, are called **mutual induction**. Their size, for given coils in a given situation, relates to the system's **mutual inductance**, M. We define this as follows.

The mutual inductance of two linked coils equals the e.m.f. induced in either when the current through the other changes by one ampere per second.

We use subscripts $_1$ and $_2$ for primary and secondary measures:

From above $\quad M = E_2 \div (\triangle I_1 / t_1)$

Also $\quad\quad\quad M = E_1 \div (\triangle I_2 / t)$

We often change these equations round like this:

$$E_1 = -M (\triangle I_2 / t);$$

and $\quad E_2 = -M (\triangle I_1 / t)$

The minus signs relate to Lenz's Law.

We do no problems on mutual inductance here. All the same you should know its unit, the henry, H.

The henry is the mutual inductance of two linked coils, in one of which one volt is induced when the current through the other changes by one ampere per second.

On what factors does M depend?

The mutual inductance of two linked coils depends on the number of turns in each. We use the symbols N_1 and N_2.
Mutual inductance depends on the sizes of their coils and their nearness to each other. It also depends on the medium linking the coils.

16.3 The voltage transformer

This is a major use of mutual induction. The two coils with which we worked in the last section are linked so that the changing magnetic field of the primary all passes through the secondary. A highly efficient energy transfer results. Its main function is to change – transform – potential differences (often called 'voltages').

In most cases the linking is an iron core. We can think of iron as an excellent conductor of magnetism – recall Fig. 15.3(d). A closed loop of iron makes a **magnetic circuit** in which there can be a good magnetic 'current'. A common design is shown in Fig. 16.5(a). Note that the core is built of thin iron sheets; these reduce energy loss. Figure 16.5 (b) gives a simpler view of the device; (c) is its circuit symbol.

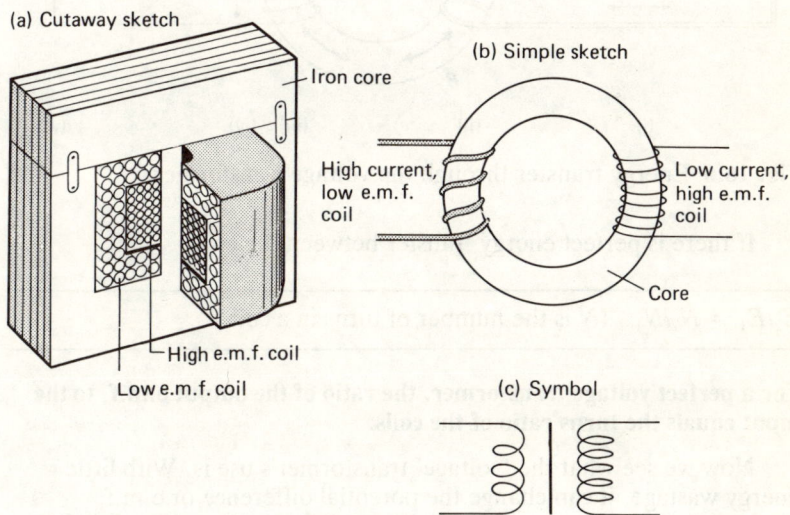

Fig. 16.5 The voltage transformer

How is the transformer supplied? In the last section, we first discussed passing d.c. through the primary coil and moving this to and from the secondary. We saw that it was much better to switch the primary current on and off.

The best way is to feed the primary with alternating current (a.c.). **Alternating current is a periodic wave form.** In the simplest cases it is like a sine wave.

Here, then, is how the transformer acts – Fig. 16.6. Note the energy changes.

(i) An alternating primary supply (E_1) causes an a.c. (I_1) in the primary coil.

202

(ii) This forms an alternating magnetic force field (*B*) along the coil axis.
(iii) The alternating field also runs along the secondary axis.
(iv) This induces an alternating e.m.f. (*E*₂) in the secondary coil.
(v) If the secondary is part of a closed circuit, an a.c. (*I*₂) will pass through it.

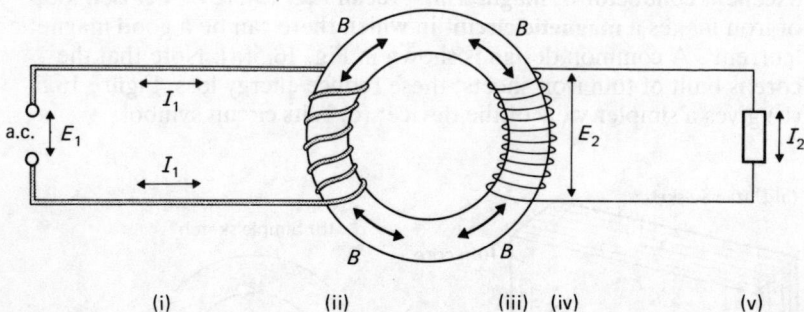

Fig. 16.6 Energy transfer through the voltage transformer

If there is perfect energy transfer between the coils:

$E_2/E_1 = N_2/N_1$ (*N* is the number of turns in a coil.)

For a perfect voltage transformer, the ratio of the output e.m.f. to the input equals the turns ratio of the coils.

Now we see what the 'voltage' transformer's use is. With little energy wastage, it can change the potential difference or e.m.f. ('voltage') supplied to a value that is more useful. The examples that follow show this.

The primary and secondary coils of a voltage transformer have 2000 and 100 turns. The supply is 240 V a.c. What is the output?

$E_2/E_1 = N_2/N_1 \rightarrow E_2 = E_1N_2/N_1 = 240 \times 100/2000$ V
$= 12$ V

What turns ratio would give 1 kV from a 50 V supply?   [194]

In this section we have often used the full name of this machine – the 'voltage transformer'. **Voltage** is widely used for potential difference or electromotive force; the voltage transformer can transform – or change – voltage, or e.m.f.

If N_2 is greater than N_1, E_2 will be more than E_1; we have a **'step up'** voltage transformer.

If N_2 is less than N_1, E_2 will be less than E_1; this describes a '**step down' voltage transformer**.

Sometimes transformers with unit turns ratio are used ($N_2 = N_1$); while not changing the e.m.f., these isolate the load from the supply and thus provide protection. They are called isolating transformers. Other transformers exist with a number of different secondary coils; they can give a number of outputs from a single supply.

How do the primary and secondary currents, I_1 and I_2, relate to each other?

Look at Fig. 16.7. Here the secondary circuit is closed – it powers some load (element) of resistance R.

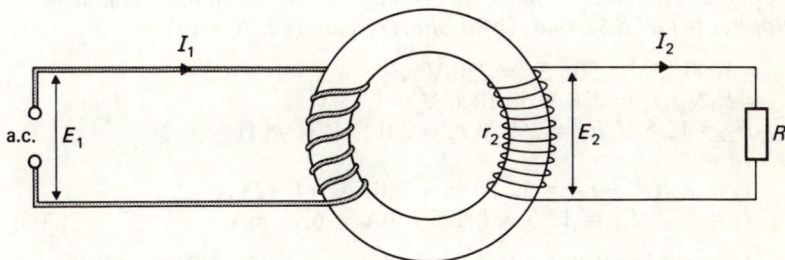

Fig. 16.7

If the secondary circuit is ohmic,

$I_2 = E_2 (R + r_2)$ (See section 14.5.)

Here r_2 is the secondary source resistance, i.e. the resistance of the secondary coil. **A transformer's output current depends on the secondary circuit resistance.**

What **power** is absorbed by the secondary circuit? **Power is the energy transferred in unit time** (section 5.3). **The power of an electric device is the product of the potential difference between its ends and the current passed.**

Here, therefore, the power absorbed by the secondary circuit is the product of the output e.m.f. and the output current.

$P_2 = E_2 I_2$

Where does the power come from? It must be supplied through the primary. If there is no power loss in the transformer,

$P_2 = P_1$

In other words $E_2 I_2 = E_1 I_1$;

or $I_1 = I_2 \times E_2/E_1$

But I_2 depends on the secondary circuit resistance ($R + r_2$).

Thus **a transformer's input current depends on the secondary circuit resistance.**

Let us say all this another way before checking with a problem.

A given transformer, supplied with e.m.f. E_1, will show e.m.f. E_2 between the ends of the secondary. If this is used to supply current I_2 to a load, I_2 depends on the load resistance. The load fixes the power absorbed by the output circuit. This cannot be more than the power taken by the input circuit. The more current taken by the load, the more must pass through the primary.

The primary and secondary of a perfect transformer have 1000 and 50 turns respectively. The resistance of the secondary coil is 2.0 Ω. The supply is 250 V a.c. What is the primary current when the secondary supplies (a) a 98 Ω load, (b) a 'short circuit' (i.e. R = 0)?

$N_1 = 1000 \quad N_2 = 50 \quad E_1 = 250 \text{ V} \quad E_2 = ?$
$E_2 = E_1 N_1/N_1 = 250 \times 50/1000 \text{ V} = 12.5 \text{ V}$
(a) $E_2 = 12.5 \text{ V} \quad E_1 = 250 \text{ V} \quad r_2 = 2.0 \text{ Ω} \quad R = 98 \text{ Ω} \quad I_2 = ?$
$I_1 = ?$
$I_2 = E_2/(R + r_2) = 12.5/(98 + 2.0) \text{ A} = 0.125 \text{ A}$
$I_1 = E_2 I_2/E_1 = 12.5 \times 0.125/250 \text{A} = 6.25 \text{ mA}$ [206]

Note that as the e.m.f. is stepped down, in effect the current is stepped up. This follows from $E_1 I_1 = E_2 I_2$.

Taking this further, we get $\quad E_2/E_1 = I_1/I_2$

But from above we have $\quad E_2/E_1 = N_2/N_1$

Therefore $\quad I_1/I_2 = N_2/N_1$

Table 16.1 relates N, E and I more closely.

Table 16.1 Currents and e.m.f.'s in a tranformer

Turns	Turns ratio	e.m.f.'s	Type	Currents
$N_2 = N_1$	$N_2/N_1 = 1$	$E_2 = E_1$	Isolating	$I_1 = I_2$
$N_2 < N_1$	$N_2/N_1 < 1$	$E_2 < E_1$	Step-down	$I_1 < I_2$
$N_2 > N_1$	$N_2/N_1 > 1$	$E_2 > E_1$	Step-up	$I_1 > I_2$

The transformers in the last three figures are step-up. This is because $N_2 > N_1$. Note that although the primary coil has fewer turns than the secondary, the wire is thicker; this is because the primary current is larger than the secondary current. (A large current needs a thicker wire than a small current – a small wire with a large current could melt.)

Do not forget that all these relationships are exact only for perfect voltage transformers. Only in these is all the energy absorbed by the primary transferred to the secondary; only then is $E_2I_2 = E_1I_1$ valid.

In practice these machines can be very efficient (see section 5.3): figures as high as 98 per cent are common. The 'lost' energy raises the temperature of coils and core; large transformers require large cooling systems to remove this wasted energy.

All the ideas met in this section are of immense value. Transformers are common in audio and video equipment, to provide the voltages needed by the various components. The national **electric power** systems ('grids') of most countries make much use of voltage transformers; this is why a.c. is supplied by the mains.

Electricity is produced in power stations at several thousand volts. One station can generate as much as a megamp at this e.m.f. To carry such a huge current to a distant city would need huge cables – and still show high loss.

Transformers are used to step the e.m.f. up to hundreds of kilovolts; the currents are then much lower, so the costs are reduced. Transformer stations and sub-stations step the e.m.f. down to safe levels. British homes use power at around 240 V; few need more than 50 A or 60 A at any one time.

A power station provides electricity at 11 kV and 100 kA. What power is generated?

$E = 11\ 000$ V $I = 100\ 000$ A $P = ?$
$P = EI = 11\ 000 \times 100\ 000$ W $= 1100$ MW

The mean power use of a house on an estate is 5.0 kW at 250 V. Five hundred such houses are supplied by a transformer in a substation. This is 95 per cent efficient; it is fed at 11 kV. (a) What is the mean current taken by one house? [254] (b) What are the input and output powers of the transformer? [318] (c) What current does the transformer take? [366]

Note: In alternating current work the power unit sometimes used is the VA (= volt-amp).

16.4 The alternator

Because the voltage transformer works only when supplied with a.c., national grid systems almost all supply this form of power. At the power stations a.c. is generated by **alternators**.

An alternator is a rotating generator of alternating e.m.f.

In practice, the design is complex. A cycle dynamo does not differ very much from the alternator in principle; a major difference is that *its* output is d.c.

A simple alternator looks just like an a.c. motor (section 15.5). Its

206

action is exactly the reverse. When supplied with a.c. the coil rotates; when the coil is turned, a.c. is produced. The first statement describes the a.c. motor; the second describes the alternator.

Figure 16.8 shows a simple alternator. Clearly, it is almost the same as Figs. 15.9 and 15.11 combined.

The shaft is turned round and round. It carries the core, the coil and the slip rings on which the brushes press. As each conductor in the coil cuts through the magnetic force field, an e.m.f. appears between its ends. These e.m.f.'s give an output current from the brushes.

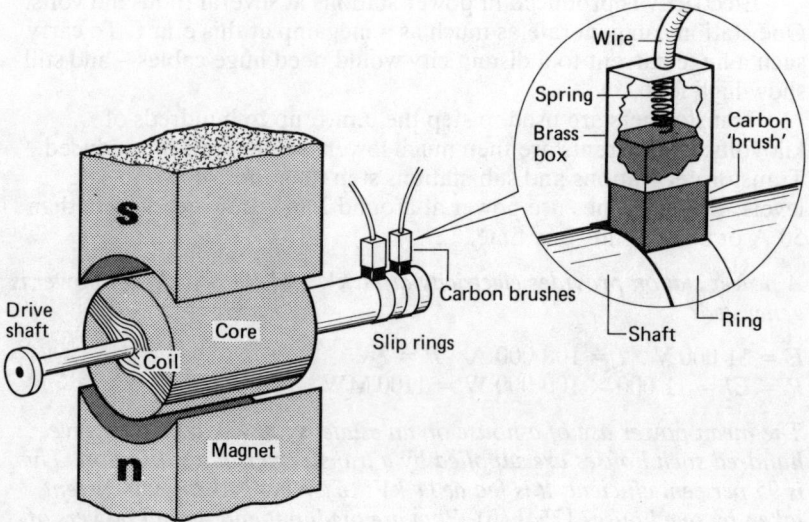

Fig. 16.8 The alternator

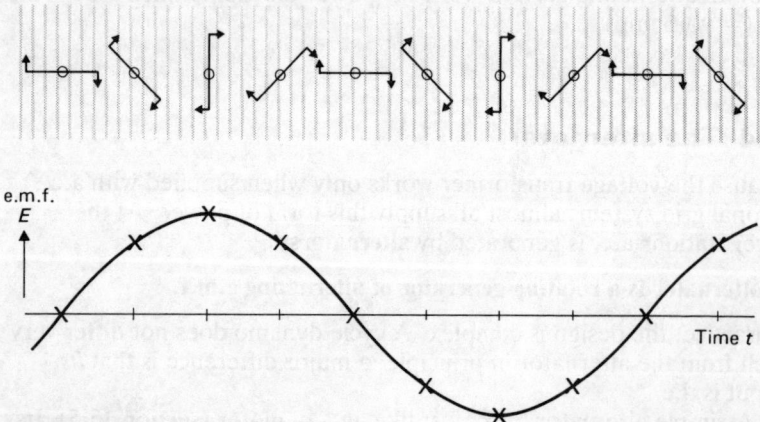

Fig. 16.9 The alternator output

At all times the laws of electromagnetic induction apply. Thus the size of the induced e.m.f. relates to the speed at which the coil cuts through the field; its direction is to oppose the motion. (This Lenz's Law opposition results from the motor effect.)

Figure 16.9 relates, in a very simple case, the e.m.f. output of the coil to the coil's position in the field.

The alternator has a sine wave form output. Applied to a load, an a.c. will pass.

The search coil technique may be used to find the strength of a magnetic force field (magnetic flux density). The action of a search coil is much the same as that of a generator.

The coil is small and flat and consists of many turns. In action it is placed at 90° to the field under test. Its ends feed any induced current to a sensitive meter. In use the coil is turned sharply through 90°. The pulse of the induced current in the circuit is recorded by the meter. This varies directly with the field strength other related factors being constant (e.g. the meter, the coil's structure, the rate of turning)

i.e. $B \propto x$ (x = the meter reading).

The technique is best used to compare the strength of an unknown field with one which is known.

Let x_1 be the meter reading for a field strength B_1 which is known

$B_1 = kx_1$ (where k is a constant)

If x_2 is the reading for an unknown field strength B_2

$B_2 = kx_2$

Hence $B_2/B_1 = x_2/x_1$

and $B_2 = B_1 x_2/x_1$

The value of an unknown field can be obtained more accurately by determining values of x for a number of known values of B. A calibration graph of x against B is plotted which may then be used to determine the value of an unknown field strength when its value of x is known.

16.5 Some more questions

1. *Sketch an 'energy web' as follows. Put electrical energy in the centre and other forms at the outside. Join as many of the forms of energy as you can with arrows labelled with a converter. Thus a 'microphone' arrow would connect sound energy to electrical energy; 'fire' would go from chemical energy to thermal energy.* [186]
2. *Describe with care how to demonstrate the laws of electromagnetic induction using the set-up of Fig. 16.4.* [372]

3. *How would you maximise the mutual inductance of two conductors?*   [210]
4. *Discuss the action of a transformer supplied with d.c.*   [435]
5. *The 5000-turn primary of a transformer takes 25 A from a 6 kV supply; the secondary has 2000 turns. Find the secondary e.m.f. and current.*   [183]
6. *A transformer supplies 15 kW. The losses are 250 W. Find the input power and the efficiency.*   [363]
7. *A 40 kW 3300/240 V transformer has 660 turns on the primary. (a) How many turns has the secondary?* [267] *(b) What are the output and input currents?*   [303]
8. *The primary of a transformer has 2000 turns. It is supplied with 240 V a.c. through a 5 A fuse. There are two secondary coils. The 1000-turn secondary supplies a 50 Ω load. What is the power of the largest load that can be supplied by the 400-turn coil?*   [195]
9. *A generator supplies 50 kW to a building. At 250 V there is a 40 V p.d. loss between the ends of the supply cables. What is the transfer efficiency?*   [295] *What p.d. would be lost 'in' the cables if the supply were at 10 kV? Find the transfer efficiency in this case.*   [438]
10 *Why does the following not give the primary current in a transformer?* $I_1 = E_1 (R + r)$; E_1 *is the supply e.m.f.,* r *is the source resistance,* R *is the coil resistance.*   [390]
11. *Explain how Lenz's Law applies to a working alternator.*   [276]
12. *Use Fig. 16.9. Explain fully why the output of the coil shown is like a sine wave.*   [322]

16.6 Objectives

When you have studied this chapter, you should be able to

(1) state changes between electrical and other forms of energy;
(2) describe simple experiments to show the laws of electromagnetic induction;
(3) state the laws of electromagnetic induction;
(4) discuss the phenomenon of mutual induction;
(5) define mutual inductance and name and define its unit;
(6) describe the action of a voltage transformer;
(7) solve simple transformer problems involving e.m.f.'s, currents, powers, turns and efficiency;
(8) explain why power is distributed at a high alternating p.d.;
(9) describe the action of a simple alternator;
(10) relate its output to how the coil rotates;
(11) describe its output as alternating e.m.f. able to supply a.c.;
(12) describe the use of the search coil to compare magnetic field strengths.

Chapter 17

Direct and alternating electricity

17.1 Measuring alternating supply

In Chapter 16 we saw that to produce alternating power is not hard. We also saw that, because the voltage transformer needs an alternating supply, this is what most power stations provide.

Mains supply is alternating; it is a sine wave form.

For many uses of electric energy it does not matter whether the supply is direct or alternating. Where it does matter, it is easy to convert a.c. to d.c. (Battery chargers and calculator mains adaptors do this, for instance.)

How do we measure alternating supply?

To describe the supply, we need *two* measures now. The first is the **amplitude**; the second is the **frequency**. We met both terms when looking at waves in Chapter 10. Here, however, the amplitude E_0 is often called the **peak value**.

The peak value (amplitude) of an alternating electromotive force is the highest value reached. Unit: volt, V

The frequency of an alternating e.m.f. is the number of cycles in unit time. Unit: hertz, Hz

---

210

In Britain, mains supply is at 50 Hz; the peak value is around 350 V. In the United States the figures are 60 Hz, 155 V.

In practice, peak value is not the best measure of an alternating e.m.f. E is at the peak E_0 for only a small fraction of each cycle. Some kind of average would be better.

What is the mean value of British mains power?   [190]

The mean value of any alternating e.m.f. is zero.

So we get no help from the mean value!

The average to use in this case is the **root mean square (r.m.s.) value**. Figure 17.1 explains this. The lightly drawn curve is of the e.m.f., E, against time. The dashed curve plots its square, E^2, against time; this has no negative values. The lightly drawn straight line is the mean of the square value, $\overline{E^2}$. The square root of this – the root mean square value of E (symbol E_{rms}) – is the heavy straight line $\sqrt{\overline{E^2}}$.

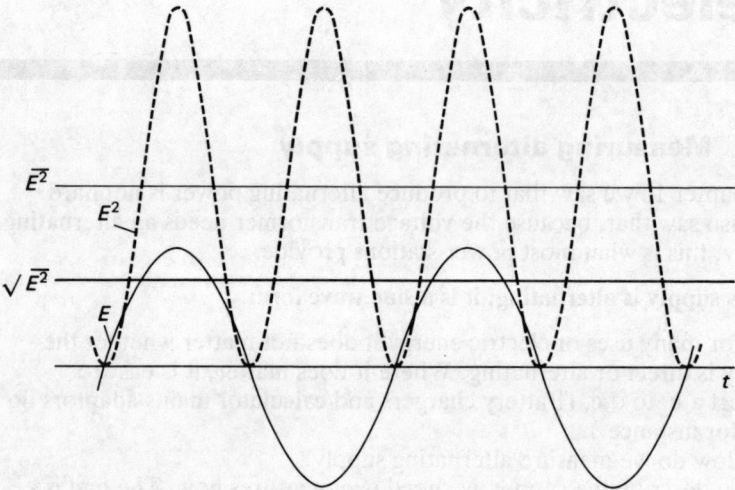

Fig. 17.1 The root mean square value of E

(Note that the square of a sine curve is also a sine curve, but not on the same axis. Its mean value is therefore not zero. This is because the square of any negative or positive value is always positive.)

$$E_{rms} = \sqrt{\overline{E^2}}$$

Compare this with the r.m.s. speed of gas particles. (See section 7.6.)

The 'average' value of an alternating quantity is the root mean square value.

This is true only if the alternating quantity is a sine wave form. In the case of an alternating current, it is also true only if the circuit is ohmic. In this book, however, we deal only with sine wave alternating quantities.

E_{rms} and E_0 must relate to each other. They do so like this.

$E_{rms} = E_0/\sqrt{2}$

The r.m.s. value of an alternating quantity is the peak value divided by $\sqrt{2}$.

Roughly what is the r.m.s. value of British mains?

$E_o = 350$ V $E_{rms} = ?$

$E_{rms} = E_o/\sqrt{2} = 350/1.414$ V $= 250$ V (approx.)

What is the r.m.s. value of United States mains?   [264]

The r.m.s. value of a transformer output is 12 V. Find the peak value.   [373]

The r.m.s. value in a major supply line is 300 kV. What is the amplitude?   [427]

The e.m.f. and frequency of an alternating supply are best found with an **oscilloscope**. (Compare with section 14.2.)

To use the oscilloscope, we should know a little about it. As shown in Fig. 17.2, it contains an electron tube rather like that in a TV set (Fig. 15.5).

A beam of high-speed electrons leaves the 'gun'. It comes to a focus on the screen; here a spot of light results from the absorbed energy. (The process is a form of fluorescence, section 12.2.) On the way to the screen, the beam passes through a set of four plates, arranged in two pairs.

Any potential difference between the plates in the first pair will deflect the beam so that the spot moves left or right on the screen. Any p.d. between the second pair of plates deflects the spot up or down. Figure 17.3 shows more clearly what can happen.

The first pair of plates moves the spot horizontally; these are the *x*-plates. The second pair is called the *y*-plates.

Not to scale; simple plan view

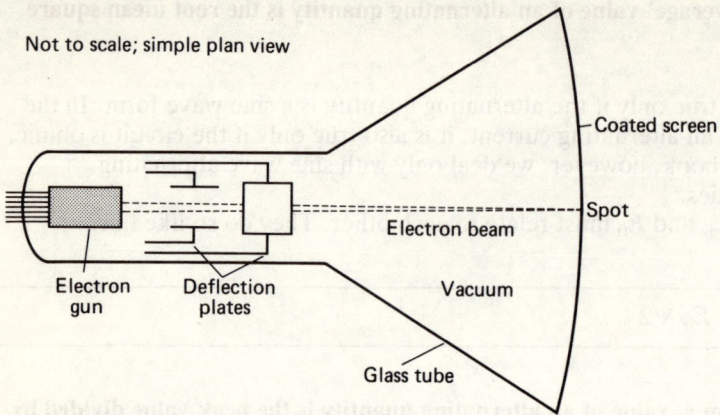

Coated screen

Spot

Electron beam

Electron gun

Deflection plates

Vacuum

Glass tube

Fig. 17.2 The oscilloscope tube

Not to scale; perspective side views

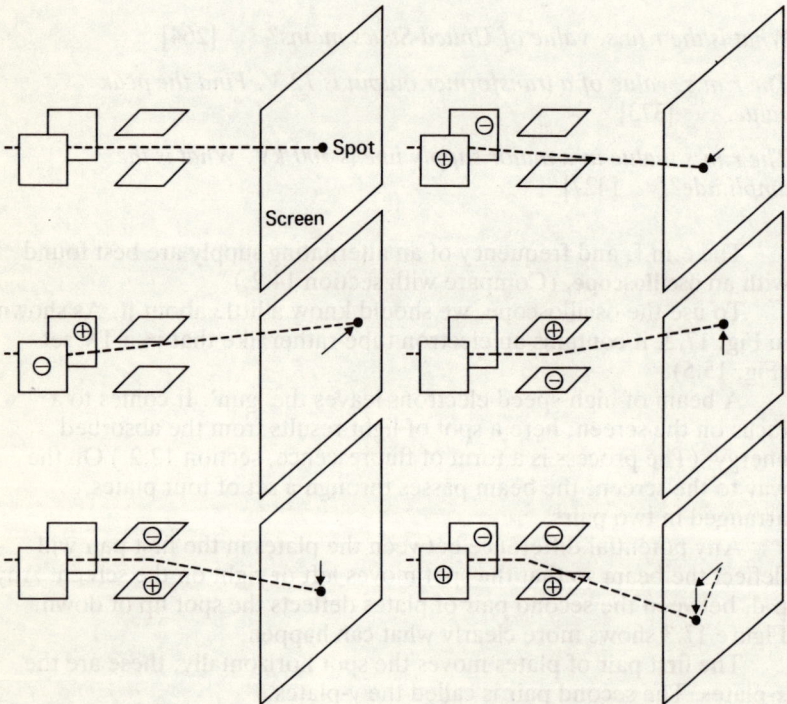

Spot

Screen

Fig. 17.3 Electrostatic deflection in an oscilloscope tube

In normal use, the signal under test is applied to the y-plates; the spot moves up and down as the p.d. between the plates changes in size and sign.

At the same time, a so-called 'sawtooth' varying potential difference – Fig. 17.4 – is applied to the x-plates. This moves the spot at a steady speed to the right; it then flies back to the left to start again. Often we call this the **time-base** signal – it can be like the time axis of a graph.

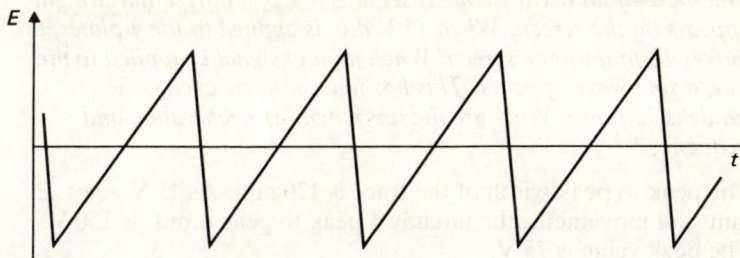

Fig. 17.4 The sawtooth wave form

Think about this case. An alternating potential difference is applied to the y-plates; its frequency is twice that of the sawtooth. How will the spot move?

In one cycle of sawtooth (a) the spot moves at a steady speed from left to right, and (b) the spot moves up and down through two cycles of s.h.m. (simple harmonic motion – Chapter 13). As a result, the *trace* on the screen is two cycles of a sine wave; see Fig. 17.5. In other words, as the timebase moves the spot once across the screen, the signal on the y-plates moves it up and down through two cycles.

We can use an oscilloscope to measure the frequency of a signal applied to the y-plates – as long as the time-base frequencies are known. The time-base frequency is changed until a small number of waves appears

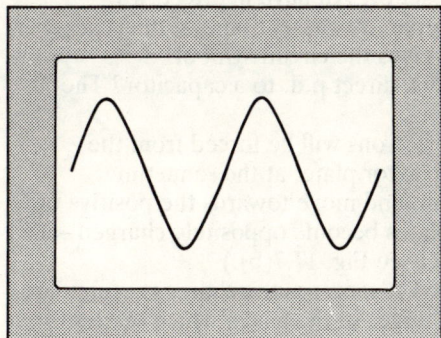

Fig. 17.5

on the screen; that frequency divided by the number of waves gives the unknown frequency. Thus in the case discussed in the last paragraph, the frequency of the signal applied to the y-plates is twice the time-base frequency.

We can use an oscilloscope to measure the peak value of a signal applied to the y-plates – as long as we know the p.d. needed to move the spot a certain distance up the screen.

The time-base of an oscilloscope is set at 200 Hz; a horizontal straight line appears on the screen. When 15 V d.c. is applied to the y-plates, the line moves 12 mm up the screen. When an a.c. signal is applied to the y-plates, a sine-wave appears. This has five complete cycles; the amplitude is 120 mm. What are the unknown (a) peak value, and (b) frequency?

(a) The peak to peak width of the trace is 120 mm. As 15 V gives 12 mm spot movement, the unknown peak to peak e.m.f. is 150 V. The peak value is 75 V.

(n) The unknown frequency is five times that of the time-base – the frequency is 1 kHz.

What is the r.m.s. value of the signal in the last question? [198]

Modern school and college oscilloscopes have y-sensitivities between about 1 mV mm^{-1} and 5 V mm^{-1}; time-base frequencies range from around 1 Hz to 1 MHz. We can therefore use them to find the frequency and peak value of a very wide range of input wave forms.

17.2 The capacitor

In essence, **the capacitor can store electric charge**. This gives it a number of uses, particularly in electronic circuits.

There are many types of capacitor. All have in effect two conducting plates; an insulator separates these. Some of the types are shown in Fig. 17.6(a) to (d); (e) gives the circuit symbol.

What will happen if we apply a direct p.d. to a capacitor? The circuit is sketched in Fig. 17.7(a).

When the switch is closed, electrons will be forced from the negative of the source to one capacitor plate; at the same time, electrons will leave the other plate and move towards the positive of the supply. Thus the capacitor plates become oppositely charged – a p.d., V_c, appears between them. (See Fig. 17.7(b).)

The p.d. between the plates, V_c, grows as the charges accumulate. However, the net force on the charges in the leads, which relates to $(V - V_c)$, will fall. The current in the circuit drops; it becomes zero when V_c reaches V.

(a) The simple air capacitor
(used for experiments)

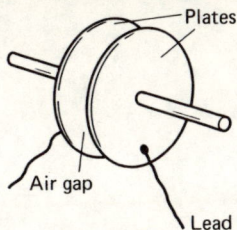

(b) The simple variable air capacitor
(used for tuning radio circuits)

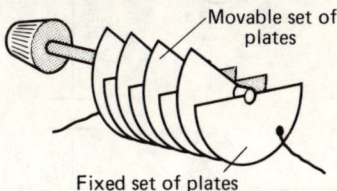

(c) Small ceramic capacitor

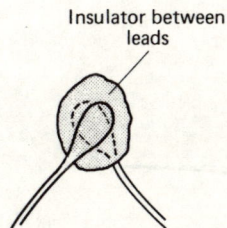

(d) Electrolytic capacitor
(contains two rolls of
foil with special paper
between)

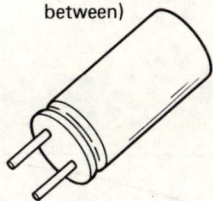

(e) Symbol

Fig. 17.6 Some capacitors

(a)

(b)

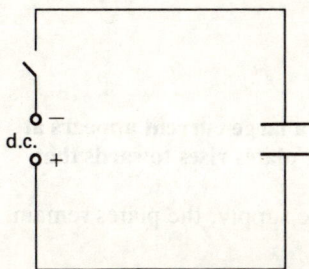

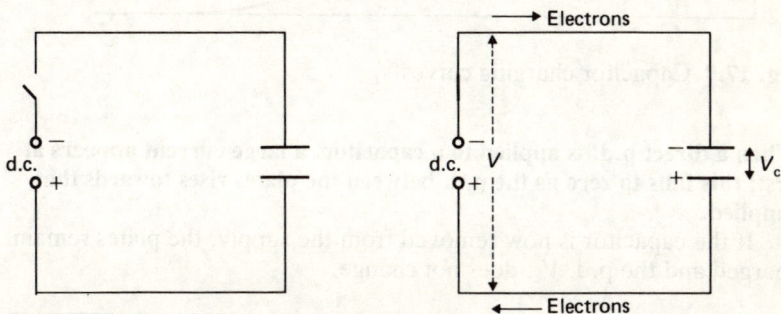

Fig. 17.7 A capacitor in a d.c. circuit

In practice, this process – charging the capacitor – is very rapid. However, a high-value resistor in the circuit will reduce the current during charging. The time to reach full charge is now much longer. This is the case in the circuit of Fig. 17.8. Here charging takes a number of seconds, so that graphs can be drawn against time of the current I in the circuit and the p.d. V_c between the plates.

The graphs obtained look like those in Fig. 17.9. The supply is switched on at time t_o.

216

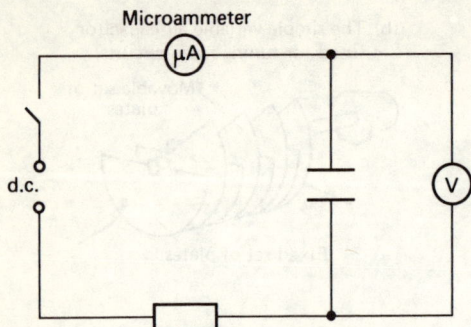

Fig. 17.8 Testing capacitor charging

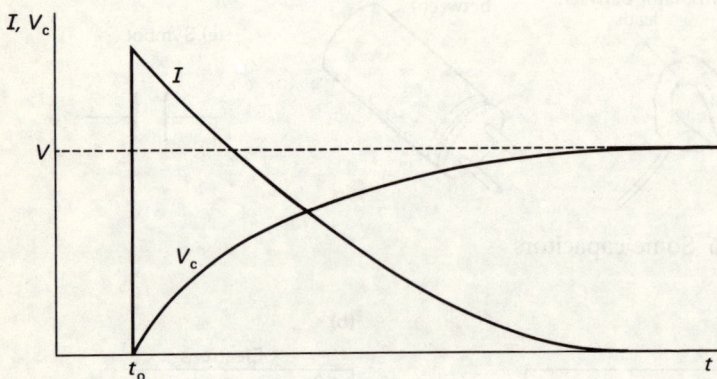

Fig. 17.9 Capacitor charging curves

When a direct p.d. is applied to a capacitor, a large current appears at first; this falls to zero as the p.d. between the plates rises towards that supplied.

If the capacitor is now removed from the supply, the plates remain charged and the p.d. V_c, does not change.

WARNING The charge in a capacitor can be shocking – handle charged capacitors with care; treat all capacitors as charged unless you *know* they are discharged.

When the leads of a charged capacitor are brought together, a spark discharge is observed. The stored electrical energy appears as 'heat', light and sound energy. **The capacitor can store energy in the form of a potential difference.**

If a capacitor is discharged through a high-value resistor, we can measure the current I and p.d. V_c during discharge. Now electrons flow from the negative plate to the positive plate because of V_c. As V falls, the current tends to zero.

The discharge curves are shown in Fig. 17.10, together with the charging curves of Fig. 17.9.

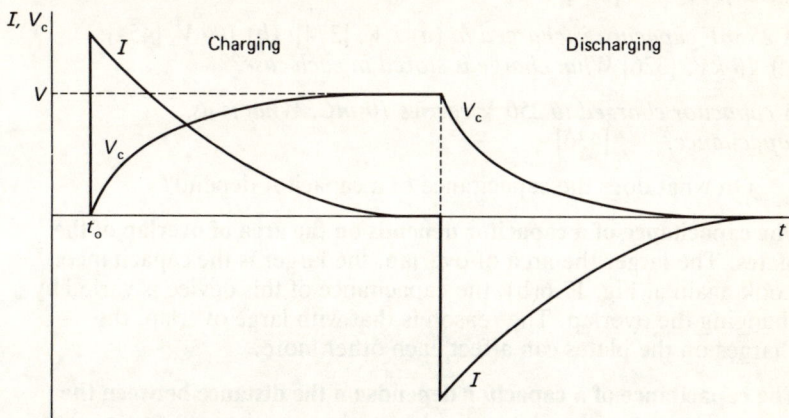

Fig. 17.10 Capacitor charge and discharge curves

How much charge, Q, can a capacitor store? Clearly this depends on V_c, the p.d. between the plates. (V_c equals the charging p.d., V.) In fact, the charge stored doubles if V is doubled.

$$Q \propto V,$$

or $Q = \text{constant} \times V$

The constant is the **capacitance** C of the capacitor.

Thus $Q = CV$.

Charge stored = capacitance × p.d. between the plates

We can rewrite this: $C = Q/V$. Unit: farad, F

The capacitance of a capacitor is the charge stored per unit p.d. between the plates.

One farad is the capacitance which stores one coulomb with one volt between the plates.

(We met the coulomb, C, the unit of charge, in section 15.2. It is the charge transferred by one ampere in one second.)

In practice, few capacitors can store much charge. Capacitances are more often quoted in microfarad, μF $(= 10^{-6}$ F$)$ or even picofarad, pF $(= 10^{-12}$ F$)$.

What charge is stored by a 100 μF capacitor at 100 V?
$C = 100 \times 10^{-6}$ F $V = 100$ V $Q = ?$
$Q = VC = 100 \times 100 \times 10^{-6}$ C $= 10^{-2}$ C.
The capacitance of an element is 50 pF. What p.d. will give it a charge of 0.1 μC?   [272]

A 25 mF capacitor is charged to (a) 1 V, [374]; (b) 100 V, [454]; (c) 10 kV. [326] What charge is stored in each case?

A capacitor charged to 250 V carries 10 mC. What is its capacitance?   [436]

On what does the capacitance of a capacitor depend?

The capacitance of a capacitor depends on the area of overlap of the plates. The larger the area of overlap, the larger is the capacitance. Look again at Fig. 17.6(b); the capacitance of this device is varied by changing the overlap. The reason is that with large overlap, the charges on the plates can affect each other more.

The capacitance of a capacitor depends on the distance between the plates. The closer the plates, the larger the capacitance. Again, with close plates, the charges affect each other more.

The capacitance of a capacitor depends on the substance between the plates. The relevant property is **relative permitivity** (ϵ_r, no unit) – the larger this is, the larger the capacitance. This is because some substances can, in effect, transmit electric forces better.

We have seen the effect of a capacitor in a direct circuit – **when a direct potential difference is applied to a capacitor, a large current appears at first; this falls to zero as the p.d. between the plates rises towards that supplied.** In other words, a capacitor cannot pass a steady direct current. We say that **a capacitor blocks d.c.**

What happens in a capacitative alternating current circuit?
When the switch is closed, an electromotive force appears in the circuit; this causes charge to move and to pile up on the plates. However, the e.m.f. direction will soon change; now charge moves the other way – the capacitor is discharged and then reverse charged.

When an alternating potential difference is applied to a capacitor, charge moves to and from the plates through the circuit – a capacitor does not block a.c.

Thus a lamp in a capacitative d.c. circuit will not light; a lamp in a capacitative a.c. circuit can light. In other words, **a capacitor blocks d.c., but not a.c.** One use of a capacitor in fact is to **filter** a.c. from d.c.

The opposition of a circuit element to a.c. is called its **reactance**, X. The reactance of a capacitor, X_c, depends on its capacitance, C, and on the supply frequency, ν.

In fact

$$X_c = 1/(2\pi\nu C)$$
Unit: ohm, Ω

A 100 μF capacitor is supplied with 50 Hz. Find its reactance.

$C = 10^{-4}$ F $\nu = 50$ Hz $X_c = ?$
$X_c = 1/(2\pi\nu C) = 1/(2\pi \times 50 \times 10^{-4})\ \Omega = 32\ \Omega$

The reactance of a capacitor at 1 kHz is 1.0 Ω. What is its capacitance?   [367]

What is the reactance of a 20 mF capacitor when supplied with (a) 10 kHz [255], (b) 10 Hz [405], (c) d.c.?   [188]

17.3 The inductor

Like the capacitor and the resistor, the inductor can absorb energy from an electric supply and thus affect the current. It, too, has a number of uses in electronics.

An inductor consists of a length of wire. For normal use this is wound into a coil and given an iron core. Often an inductor may look very like a voltage transformer (Fig. 16.5); however, it has one coil, not two, so has two leads rather than four. Because of the usual design, inductors are often called **coils**.

What will happen if we apply a direct potential difference to an inductor (coil)? Look at Fig. 17.11; note the inductor's circuit symbol.

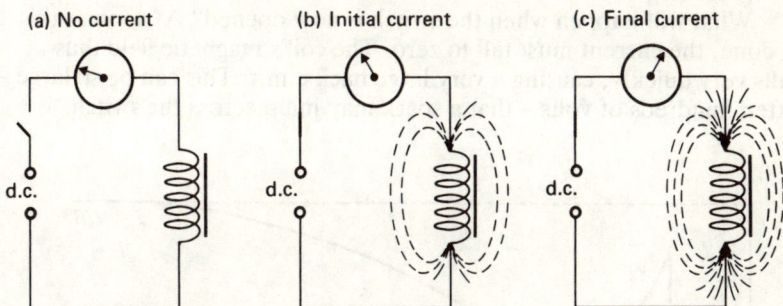

(a) No current   (b) Initial current   (c) Final current

Fig. 17.11 An inductor (coil) in a d.c. circuit

When the switch is closed, a current will grow from zero in the coil, (b).

At the same time, magnetic field appears round the coil. This field grows as the current grows.

Thus we have a conductor in a changing magnetic field. This is the condition for **electromagnetic induction** (section 16.1). Therefore an e.m.f. appears in the coil to oppose the cause – to oppose the growing current. This e.m.f. is called a **back e.m.f.**; the effect is **self-induction** (compare with section 16.2).

As a result, **when a direct potential difference is applied to a coil, the current grows only slowly.**

Of course, once the current reaches its final value (V/R), the back e.m.f. becomes zero. This is the case in Fig. 17.11(c).

We can show the effect of self-inductance with the circuit of Fig. 17.12.

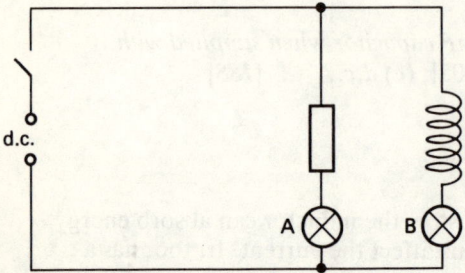

Fig. 17.12 To show self-inductance

When the switch is closed, lamp A lights at once. Lamp B comes on more slowly. However (if the resistor and the coil have the same resistance) the lamps will have the same final brightness.

If we plot the current I in a coil and the back e.m.f. V_L after the switch is closed at time t_0, we obtain the graphs of Fig. 17.13. (The use of the subscript $_L$ for a coil becomes clear later.)

What will happen when the switch is now opened? As soon as this is done, the current must fall to zero. The coil's magnetic field thus falls very quickly, causing a very large back e.m.f. This can be so large – often hundreds of volts – that a spark may jump across the switch.

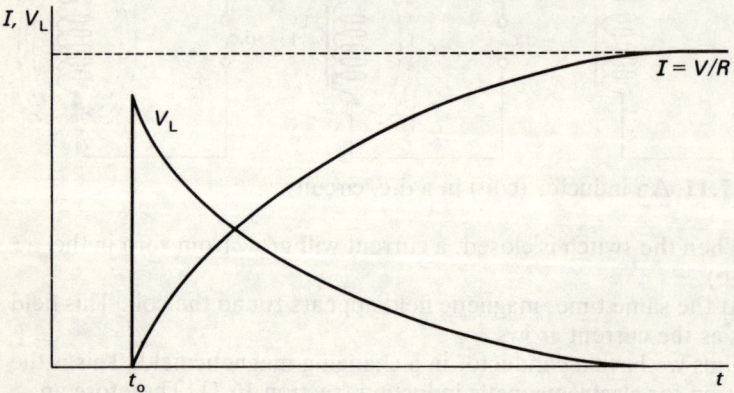

Fig. 17.13 Curves for an inductive d.c. circuit

You may have noticed this when switching off a fluorescent lamp – the spark can be heard if the switch is not very good. The effect is used, for instance, in the ignition coil of an engine – this can make very good sparks from a car's 12 V supply.

Now let us think about an alternating inductive circuit. With an alternating supply the current is changing all the time. Therefore the coil's magnetic field is changing all the time. Therefore there is a back e.m.f. all the time. Therefore the a.c. is opposed all the time.

In other words, like a capacitor, a coil has some opposition to a.c. The alternating current is less than we would expect from the coil's resistance alone. The opposition depends on the coil's *inductance*.

The reactance of a coil depends on its **inductance**. (It is usual to use this word rather than '*self-inductance*'.)

We can define the inductance L of a coil as in section 16.2, using the induced e.m.f. V_L

$$V_L = L \, \triangle \, I/t$$

The e.m.f. in a coil = inductance × current change ÷ time taken

---

The minus sign some people use here relates to Lenz's Law; it reminds us that V_L is the *back* e.m.f.

Changing this round gives

$$L = V_L \div \triangle \, I/t$$

The inductance of a coil is the back e.m.f. induced when the current in it changes by one ampere per second.

The unit of inductance is the **henry**, H, as for mutual inductance (section 16.2).

The henry is the inductance of a coil in which one volt is induced when the current in it changes by one ampere per second.

The reactance of a coil X_L depends on its inductance, L, and on the supply frequency, v.

In fact $X_L = 2 \pi v L$ Unit: ohm, Ω

A 100 mH coil is supplied at 50 Hz. Find its reactance.

$L = 0.1$ H $v = 50$ Hz $X_L = ?$
$X_L = 2 \pi v L = 2 \pi \times 50 \times 0.1 \, \Omega = 31 \, \Omega$

The reactance of a coil at 1 kHz is 1 k Ω. What is its inductance?   [229]

What is the reactance of a 20 H coil when supplied with (a) 10 kHz a.c., (b) 10 Hz a.c., (c) d.c.?   [207]

It is useful to compare the effects of capacitors, inductors and resistors in direct and alternating circuits. This is done in Table 17.1, where l.f. means low frequency, and m.f. and h.f. refer to medium and high frequencies respectively.

Table 17.1 Elements in d.c. and a.c. circuits

Element	Capacitor	Inductor	Resistor
Symbol	⊣⊢	⬚⬚⬚	—▭—
Specific property	capacitance, C	inductance, L	resistance, R
Unit of property	farad, F	henry, H	ohm, Ω
Effect on steady d.c.	infinite opposition	no opposition	same opposition
Effect on l.f. a.c.	high opposition	low opposition	same opposition
Effect on m.f. a.c.	some opposition	some opposition	same opposition
Effect on h.f. a.c.	low opposition	high opposition	same opposition
Reactance	$X_C = 1/(2\pi v L)$	$X_L = 2\pi v L$	$X_R = R$

Capacitors, inductors and resistors are called **passive circuit elements** (sometimes **reactors**). As far as we are concerned here, their only effect is to oppose charge flow in different ways. The whole science of electronics is based on their use, in conjunction with that of active elements like diodes and transistors.

17.4 Some more questions

1. *Refer to Fig. 17.1. Show that, for a pure sine wave,*
   $E_{rms} = E_0/\sqrt{2}$.   [331]
2. *Draw with care curves like those in Fig. 17.1, with E_o equal to*
   (a) 0.50 V, [383]; (b) 1.0 V, [224]; (c) 1.5 V. [208] Find from your
   curves the r.m.s. e.m.f. in each case.
3. *Supplied with 120 V a.c., a transformer gives 12 V and 240 V*
   outputs. What are the peak values of these three e.m.f.'s?   [209]
4. *The peak-to-peak value of a signal is 150 μV. What is its r.m.s.*
   value?   [288]
5. *An unknown signal is applied to the y-plates of an oscilloscope.*
   When the time-base is set at 20 Hz, the trace on the screen is as
   shown in Fig. 17.14. Find the signal's (a) frequency [158],
   (b) r.m.s. value [258]. The y-sensitivity is 4.0 mm V^{-1}.
6. *Explain with care the form of the curves in Fig. 17.10.*   [442]
7. *Design a circuit with one source, one high-value resistor and one*
   high-value capacitor, which will allow you to obtain the curves of
   Fig. 17.10. Your other elements may include a double throw switch
   like that in Fig. 17.15.   [236]

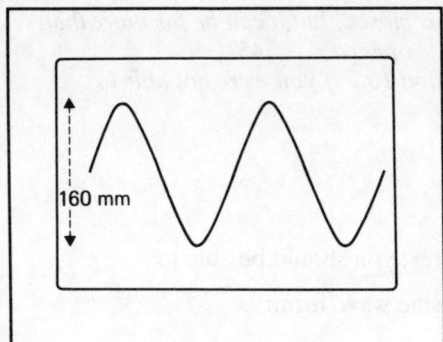

Fig. 17.14

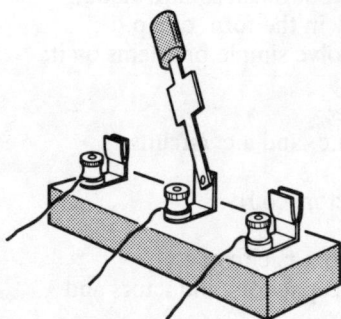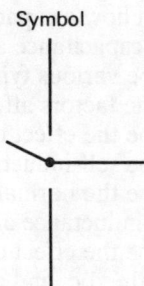

Symbol

Fig. 17.15

8. *Copy and complete Table 17.2. It concerns a number of capacitors.*

Table 17.2

	Capacitance	Applied p.d.	Charge stored	
(a)	500 pF	2 kV		
(b)	50 mF	250 V		[164]
(c)	100 μF		2 C	
(d)	0.02 F		100 μC	[178]
(e)		250 mV	50 μC	
(f)		5 kV	2 mC	[212]

9. *Find the reactances of*
   (a) a 100 Ω resistor at (i) 10 Hz, (ii) 10 MHz; [173]
   (b) a 100 μF capacitor at (i) 10 Hz, (ii) 10MHz; [197]
   (c) a 100 mH inductor at (i) 10 Hz, (ii) 10MHz. [307]
10. *Explain with care why the back e.m.f. in a coil cannot be more than*

that supplied when the switch is closed, but it can be far more than that supplied when the switch is opened.   [452]

11. *Attempt question 10 from section 16.5 if you were not able to answer it before.*   [403]

17.5 Objectives

When you have studied this chapter, you should be able to

(1) state that mains supply is of sine wave form;
(2) explain why this is so;
(3) define the frequency of an alternating supply;
(4) define and relate for an alternating supply, peak-to-peak value, peak value, mean square value, and root mean square value;
(5) explain how a capacitor stores energy in the form of a p.d.;
(6) define capacitance and its unit, and solve simple problems on it;
(7) describe various types of capacitor;
(8) state the factors affecting capacitance;
(9) describe the effect of a capacitor in d.c. and a.c. circuits;
(10) describe self-induction;
(11) describe the normal form of an inductor (coil);
(12) define inductance and its unit;
(13) describe the effect of a coil in d.c. and a.c. circuits;
(14) relate the d.c. and a.c. behaviour of capacitors, inductors and resistors.

Chapter 18

Electric force

18.1 The electric force field

The concept of electric **charge** was first raised in Chapter 14; it has been used a number of times in later chapters.

Now we develop the idea of **electric forces**: forces on charges. Here is how we described these at the start of Chapter 14.

Like charges repel each other; unlike charges attract each other.

In one respect electric forces (forces between charges) are like magnetic forces (forces between poles – section 15.1) and gravitational forces (forces between masses) – they can act without contact.

Thus an **electret** (like a rubbed plastics rod) can attract small pieces of matter without contact – just as a magnet can pick up nails, just as the Earth pulls an apple down from a tree.

Nuclear forces i.e. those between nuclear particles (section 19.2) are also like this.

No one knows how any of these forces can act at a distance without contact; of course there are many theories. Instead we *invent* the idea of a **force field** to describe what we observe. Force fields do not explain; they simply describe. (Perhaps the human 'mind' is the same kind of concept!) Note that force fields are often called just 'fields'.

A (force) field is a volume of space in which certain forces can be observed. We define each type of (force) field as follows.

A gravitational (force) field is a region of space in which gravitational forces are observed.

A magnetic (force) field is a region of space in which magnetic forces are observed.

An electric (force) field is a region of space in which electric forces are observed.

A nuclear (force) field is a region of space in which nuclear forces are observed.

In this chapter we discuss electric fields and electric forces in some detail.

An electric field is a region of space in which electric forces are observed. As electric forces act only on electric charges, we use electric charges to test for the presence of electric fields.

One approach uses a small polystyrene foam ball (very light), coated with graphite (a conductor), hung on a nylon thread (an insulator). We charge the ball by touching it with a rubbed ebonite rod (making it negative), or with a rubbed acetate rod (making it positive).

If the ball hangs freely no electric field is acting. If it moves to one side or the other, i.e. it shows a force, it must be in a field.

The size of an electric force field is the **electric field strength**, E.

The electric field strength at a point is the force on unit charge at that point.

We can write

$E = F/Q$ Unit: newton per coulomb, NC^{-1}

(We described gravitational field strength in rather the same way in section 2.2 – $g = W/m$; unit: newton per kilogram, $N\ kg^{-1}$.)

The force on a charge is 5 mN in a 20 NC^{-1} field. What is the charge?

$F = 5 \times 10^{-3}\ N\quad E = 20\ NC^{-1}\quad Q = ?$
$E = F/Q \rightarrow Q = F/E = 5 \times 10^{-3}/20\ C = 2.5 \times 10^{-4}\ C$

This charge is now placed where the electric field strength is 0.4 kNC^{-1}. What force acts on it?

$Q = 2.5 \times 10^{-4}\ C\quad E = 400\ NC^{-1}\quad F = ?$
$F = EQ = 400 \times 2.5 \times 10^{-4}\ N = 0.1\ N$

The charge is next doubled and moved. The force on it becomes 1 kN. What is the new electric field strength? [457]

Note that electric field strength is a vector. (All field strengths are vectors, as force is a vector.) **The direction of the electric field at a point is the direction of the force on a positive charge at that point.**

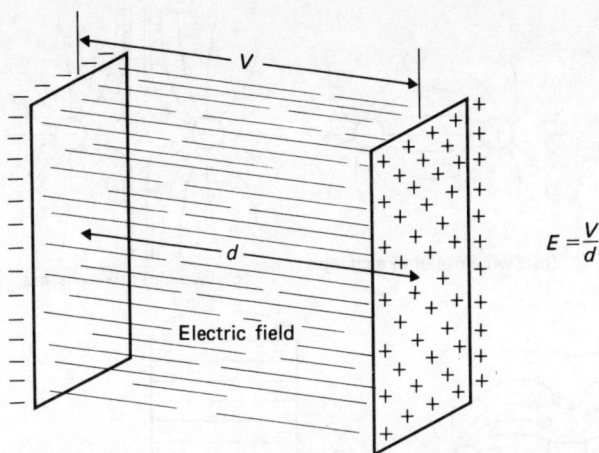

Fig. 18.1 To define electric field strength

We can also define the unit of electric field strength as the volt per metre, $V\ m^{-1}$. This relates to a second expression for E, as between the two plates in Fig. 18.1.

$E = V/d$ Unit: volt per metre, $V\ m^{-1}$

Electric field strength = the p.d. between two points/ the distance between them

---

The potential difference between two plates 100 mm apart is 5 kV. What is the strength of the field between them?   [439]

The field between two plates is 100 $\mu N\ C^{-1}$. If the plates are 5 m apart, what is the p.d. between them?   [404]

When we discussed magnetic force fields in Chapter 15, we looked at their shapes, in two and three dimensions. (See Figs. 15.2 and 15.3.) The shapes of electric fields are much the same; we can see this in Fig. 18.2. These are sections in two dimensions.

As with magnetic field pictures, we use lines to describe the form of electric fields. These lines are not real; however they show (by their direction) the field direction, and (by their closeness) the field strength.

18.2 What is potential difference?

In the same way as gravitational potential energy relates to gravitational force (weight) – section 5.2 – electric potential difference relates to electric force.

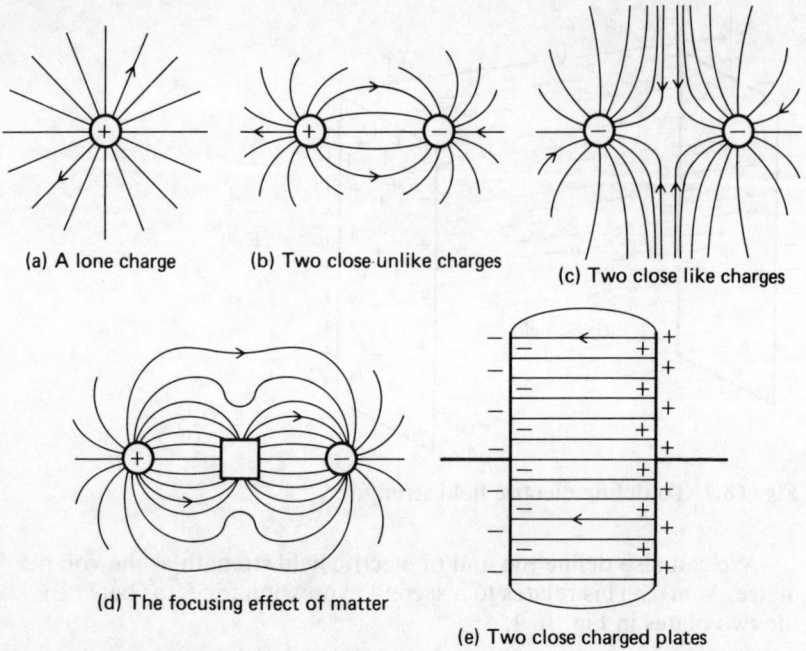

(a) A lone charge   (b) Two close unlike charges

(c) Two close like charges

(d) The focusing effect of matter

(e) Two close charged plates

Fig. 18.2 Some electric fields

Electric potential difference (p.d.) was mentioned first in section 14.1. We said that it measures the difference in charge level between two points. We then said that the free charges in a wire joining the points would drift, to make a current. Those ideas were perhaps a little too simple.

Refer to Fig. 18.3, a new version of Fig. 14.1.

If there is a potential difference between two points, there must be an electric field between them. (We can explain this on the basis of charge levels, or by using $E = V/d$.) An electric field is a space where electric charges will experience forces. A metal wire contains many free charges – the so-called electron gas (section 19.1). These will tend to accelerate in a field, and set up a current (a flow of charge).

This acceleration requires **energy**. Strictly we define potential difference in terms of energy.

The potential difference between two points is the energy involved in moving unit charge between them. (We could say 'work done' rather than 'energy involved'; as we saw in Chapter 5, these mean the same.)

To show that this statement about potential difference is valid, we use three relations we have met before.

$P = W/t$ (section 5.3) $P = VI$ (section 5.3) $Q = It$ (section 15.2)

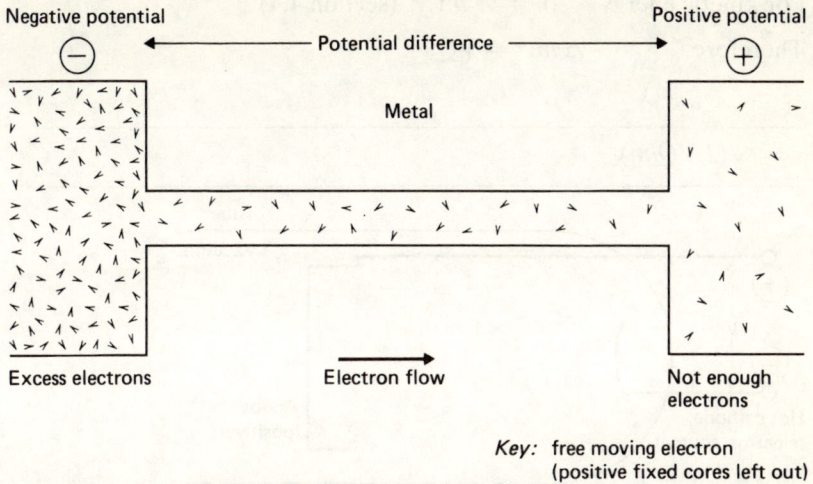

Fig. 18.3 Potential difference

From these we get $W = VIt$

and then $W = VQ$

or

$V = W/Q$

This is a statement of how we define potential difference – **the energy involved in moving unit charge between the points concerned.**

In the case of a current in a metal, the free charges 'drift' along rather than having a steady acceleration. Their motion remains almost random, as they collide with each other and with the fixed cores. The electron's mean *speed* may be hundreds of metres per second; the mean *velocity* is only a few millimetres a second.

If there is nothing in the way, a free charge would accelerate all the way across a potential difference. The energy gained from the field would then all appear as kinetic energy (section 4.4), instead of the temperature rise that results from the collisions.

Electrons are accelerated in this way in many modern 'electron tubes'. In each case there is a vacuum in the tube to reduce energy loss by collision. The electron gun of an oscilloscope, the X-ray tube, the old vacuum tubes – in each, electrons are accelerated in a vacuum by a field.

The potential difference, V, between the negative and positive electrodes (**cathode** and **anode**) may be several hundred volts. What speed will an electron gain as it crosses that gap?

We have $V = W/Q$

This gives $W = VQ$

230

For kinetic energy $W = \frac{1}{2} mv^2$ (section 4.4)

Therefore $\frac{1}{2} mv^2 = VQ$

or

$v = \sqrt{(2\,VQ/m)}$

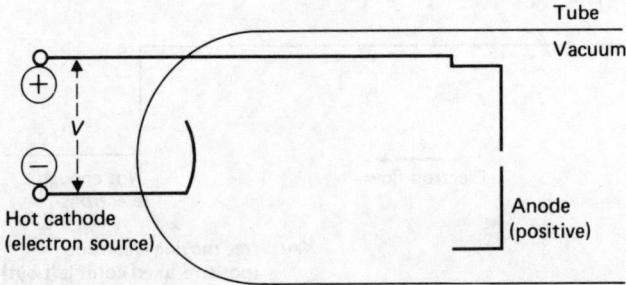

Fig. 18.4 An electron gun

A particle of total mass 10^{-20} kg carries 1000 electrons. What will be its kinetic energy if it accelerates from rest through 100 kV? Find the final speed. The electron charge is 1.6×10^{-19} C;
$m = 10^{-20}$ kg $Q = 10^3 \times 1.6 \times 10^{-19}$C $V = 10^5$V $W = ?$ $v = ?$
$W = VQ = 10^5 \times 10^3 \times 1.6 \times 10^{-19}$J $= 1.6 \times 10^{-11}$J
$W = \frac{1}{2} m v^2 \rightarrow v = \sqrt{2W/m} = (3.2 \times 10^{-11}/10^{-20})$ m s^{-1} = 57 km s^{-1}

In an electron gun, the cathode-anode p.d. is 750 V. At what speed will electrons leave the gun? Mass is 10^{-30} kg.

$V = 750$ V $Q = 1.6 \times 10^{-19}$C $m = 10^{-30}$ kg $v = ?$
$v = \sqrt{(2VQ/m)} = \sqrt{(2 \times 750 \times 1.6 \times 10^{-19}/10^{-30})}$ m s^{-1}
    $= 15.5$ Mm s^{-1}

Use the same data to find the speed gained by an electron in a triode valve. The p.d. here is 120 V.   [364]

18.3 The electron

It seems that the electron, a tiny piece of negative charge, is one of the basic particles of which all matter is made. (We say more about this in the next chapter.)

We have just seen that its charge is -1.6×10^{-19} C and its mass is about 10^{-30} kg. How have such tiny measures been made?

After electrons were discovered at the end of the last century, many experiments were done. Their properties and behaviour were assessed; people tried to find out just what they were. That they were in fact particles was not clear at first; an early name, which is still used, was '**cathode rays**' – radiation from the negative electrode.

In this chapter we discuss two of those many experiments. In the one, first carried out by J.J. Thomson in 1897, the electronic **specific charge** is measured. This is the ratio Q_e/m_e. In the second (by Millikan in 1909), the electronic mass m_e is found. Thus we can work out the electron charge Q_e (sometimes called e).

A modern form of Thomson's experiment uses the apparatus shown in Fig. 18.5.

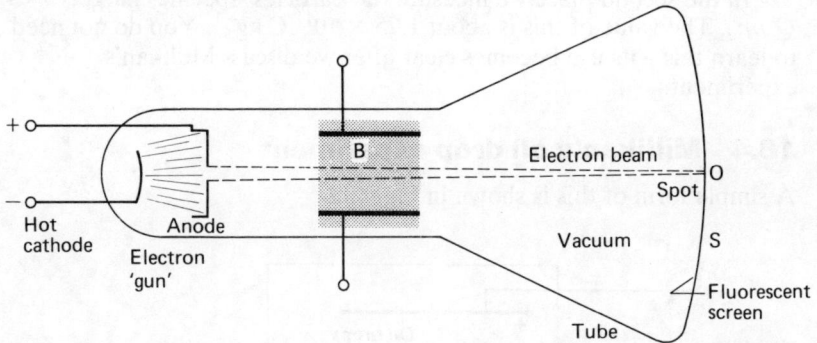

Fig. 18.5 To find Q_e/m_e

First the electron beam is allowed to pass straight through; O is marked on the screen.

Next a known magnetic field B is applied. We show it shaded in Fig. 18.5. Its direction is 'into the paper'; use of the left-hand motor rule – Fig. 15.4 – shows that the beam will be turned so that the spot moves down the screen, to S say.

Lastly a potential difference is applied between the plates, the upper one being positive. The p.d. is changed until at value V, the spot returns to O.

We thus balance the motor force ($F = BQ_ev$, section 15.2) against the electric force ($F = Q_eE$, section 18.1):

$$BQ_ev = Q_eE$$

or $\qquad v = E/B$

When B alone acts, each electron is forced to move along the arc of a circle; the motor force is the centripetal force ($F = mv^2/r$, section 3.3):

$$BQ_ev = m_ev^2/r$$

The specific charge is given by

$$Q_e/m_e = v/rB,$$

or

$$Q_e/m_e = E/rB^2$$

The field strengths E and B can be measured or calculated; r, the radius of the electrons' path through B, is found by geometry.

The experiment is useful for two reasons. Firstly we find that v, the electron speed, is far less than the speed of light. Thus 'cathode rays' are not electromagnetic waves.

In the second place we measure the particles' specific charge, Q_e/m_e. The value of this is about 1.75×10^{11} C kg^{-1}. You do not need to learn this – its use becomes clear after we discuss Millikan's experiment.

18.4 Millikan's oil drop experiment

A simple form of this is shown in Fig. 18.6.

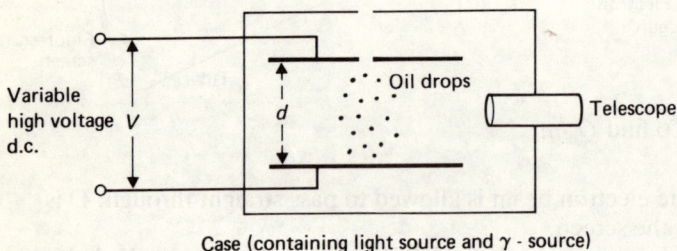

Case (containing light source and γ - source)

Fig. 18.6 To find Q_e

Some tiny drops are sprayed into the space between the plates. A burst of gamma-rays from a radioactive source will charge some of them negatively. We then adjust the field between the plates so that the drop viewed stays still – it neither falls because of its weight nor rises because of the upward electric force on it.

We thus balance these two forces.

Then $mg = EQ_c$.

Here m and Q_c are respectively the mass and charge of the drop.

This gives $Q_c = mg/E$.

The mass of the drop can be found from its volume and density – $\frac{4}{3} \pi r^3 \rho$. The field E is V/d. We know the value of g.

Thus Q_c, the charge of the drop, is found. This is in fact the charge of some number of electrons. Each time the experiment is done, we find that Q_c is some multiple of 1.6×10^{-19} C.

We conclude that the electron charge, Q_e, is 1.6×10^{-19} C. So far no smaller (non-zero) value of charge has been confirmed.

The Thomson experiment gave the specific charge, Q_e/m_e, 1.75×10^{11} C kg^{-1}.

What is the mass of an electron?   [339]

These days simple versions of the Thomson and Millikan experiments are cheap and easy to use. With them, even in a small college laboratory, we can measure the electron's tiny mass and charge.

18.5   Some more questions

1. *Referring as required to Fig. 18.2(d), explain with care how an electret (like a rubbed plastics rod) can pick up a piece of paper.*   [355]
2. *Study Fig. 18.7. It shows a light uncharged ball; this hangs between two plates oppositely charged to a high p.d. The ball is tapped so that it touches one plate. What will happen? Explain fully; give attention to electric field and charge.*   [375]

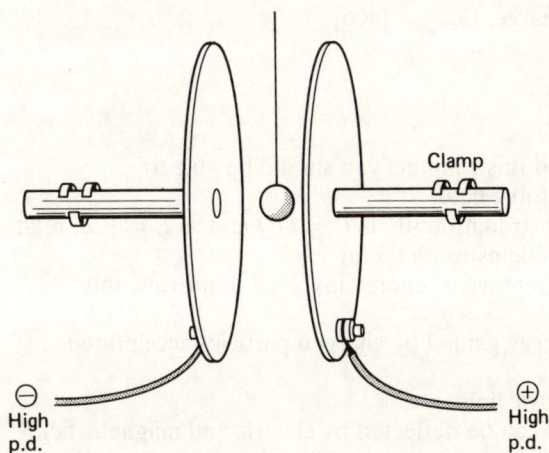

Fig. 18.7

3. *A 20 mC charge is placed in turn in fields of 1 Vm^{-1} (NC^{-1}), 1 kVm^{-1}; 1 MVm^{-1}. What is the force on the charge in each case?*   [394]
4 *The ball in Fig. 18.7 carries a 1 μC charge. 6 kV is applied between the plates, which are 100 mm apart. If the ball's mass is 120 g, with what acceleration will it start to move?*   [443]

234

5. *The same ball, with the same charge, is found to accelerate at $10 \, ms^{-2}$ in an electric field. How strong is the field?* [464]

6. *The plates of an air capacitor are 0.1 mm apart. What is the field between them when the p.d. applied is (a) 1 V, (b) 1 kV, (c) 5 kV?* [487]

7. *Sketch the field between three equal charges, two negative and one positive, at the corners of an equilateral triangle.* [495]

8. *The current in a circuit section is 3 A, the p.d. between the ends being 4 V. Find (a) the resistance of the section* [444]; *(b) the charge passed in (i) a second, (ii) a minute* [386] *(c) the energy transferred in (i) a second, (ii) a minute* [479].

9. *The anode-cathode p.d. in an X-ray tube is 50 kV; the electrodes are 200 mm apart. What are (a) the field in the gap* [357]; *(b) the acceleration of an electron in the gap* [211]; *(c) the speed at which an electron hits the anode* [493]? *The electron mass and charge are 10^{-30} kg and 1.6×10^{-19} C.*

10. *The last part of the last problem could be solved either with the equation at the end of section 18.2 or with the equations of motion. Show why the two methods give the same answer.* [377]

11. *Explain in your own words, with your own sketch, an experiment to find the electron specific charge, Q_e/m_e.* [397]

12. *Explain in your own words, with your own sketch, an experiment to find the electron charge, Q_e.* [406]

18.6 Objectives

When you have studied this chapter, you should be able to
(1) define any kind of force field;
(2) define electric field strength both as $E = F/Q$ and as $E = V/d$, relate these and solve problems using them;
(3) define potential difference as energy involved in moving unit charge;
(4) find the kinetic energy gained by charged particles accelerated through a p.d.;
(5) sketch various electric fields;
(6) state that electrons can be deflected by electric and magnetic fields;
(7) describe and explain an experiment to find the electron specific charge Q_e/m_e;
(8) describe and explain an oil drop experiment to find the electron charge.

Chapter 19

The atom

19.1 Ionisation

In this final chapter we relate the concepts about 'the atom' we have met before; we also discuss its structure in more detail.

An atom is the smallest part of an element that can exist alone. (See section 6.1.) The hundred or so elements are the basic substances from which all matter is made. No doubt you know the difference between elements and compounds; Table 19.1 gives examples of each, with the symbols used in science.

Table 19.1 Elements and compounds

Elements (symbols)		Compounds (symbols)
Hydrogen	(H)	Water (H_2O)
Helium	(He)	Carbon dioxide (CO_2)
Carbon	(C)	Iron oxide (Fe_2O_3)
Oxygen	(O)	Hydrogen chloride (HCl)
Iron	(Fe)	Copper chloride ($CuCl_2$)
Copper	(Cu)	Methane (CH_4)
Neon	(Ne)	Methyl chloride (CH_3Cl)
Uranium	(U)	Polyisoprene (rubber) ($C_{15}H_{24}$)$_n$
Chlorine	(Cl)	

In fact it is too simple to talk of elements and compounds like this – but the concepts are useful.

Each atom has a small central nucleus; this carries a positive electric charge. A cloud of negative charge surrounds the nucleus, carried by negative electrons.

Figure 19.1 shows the kind of model ('picture') of an atom we now have.

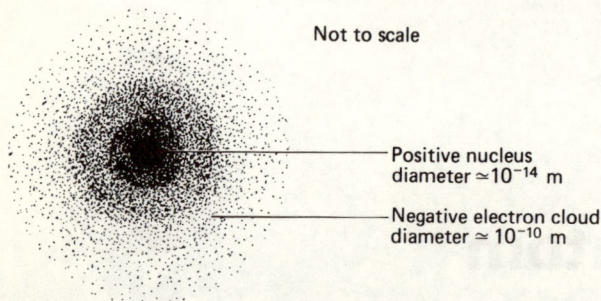

Not to scale

Positive nucleus
diameter $\simeq 10^{-14}$ m

Negative electron cloud
diameter $\simeq 10^{-10}$ m

Fig. 19.1 A simple model of the atom

In normal matter the total positive charge of all the nuclei equals the total negative charge of all the electrons. Indeed, this is true to some extent of each atom – the positive nuclear charge equals the negative charge of the electron cloud. **Atoms tend to be neutral.** However, in a **metal** one or more of the electrons from each atom do not stay in the cloud; they wander throughout the substance in a **free electron 'gas'**. So metals conduct electricity well – a current of these free charges is easy to set up.

If an atom does not have the right number of negative electrons in its cloud to balance the positive nuclear charge, it is called an **ion**.

An ion has too few or too many electrons, giving it a net positive or negative charge. These ions are called, respectively, **cations** and **anions**. (In an electric field, they tend to move towards the cathode (−) and the anode (+) respectively.)

Ionisation involves adding electrons to, or taking electrons from, a neutral atom.

Ionisation needs energy. An electron in the cloud needs some energy to make it free; an outside electron must be forced into the electron cloud. In either case an electron is moved against an electric force.

In section 12.5 we found that the electrons in the cloud have different energies. Each atom has a series of **energy levels** which relates to how the electrons can be arranged. We saw, too, that atoms need energy to move electrons to higher levels; they release energy if electrons 'fall ' to lower levels.

The same kind of thinking is involved with ionisation; atoms need energy for this to happen.

The ionisation energy of an atom is the energy needed to cause an electron to escape from the cloud. (We shall not discuss the formation of negative ions here.)

Look again at an energy level diagram – Fig. 19.2.

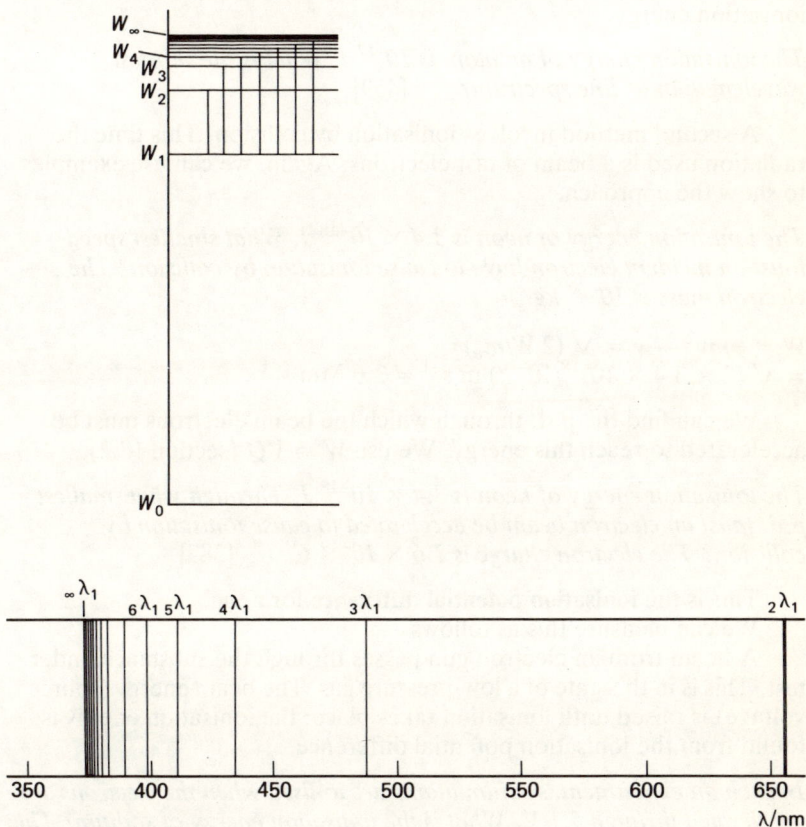

Fig. 19.2 The energy levels of hydrogen gas

Note: the scale is linear

Here the level W_∞ is the ionisation energy level for hydrogen – the highest possible.

Radiation of frequency 3.3×10^{15} Hz is absorbed when hydrogen gas is ionised. What is the ionisation energy? The Planck constant, h, is 6.6×10^{-34} J s.

$v = 3.3 \times 10^{15}$ Hz $h = 6.6 \times 10^{-34}$ J s $W = ?$
$W = hv$ (section 12.5)
$= 6.6 \times 10^{-34} \times 3.3 \times 10^{15}$ J $= 2.2 \times 10^{-18}$ J

Neutral helium atoms have two electrons. The energies to remove them are 3.9×10^{-18} J and 8.7×10^{-18} J. Find the wavelengths needed to cause these ionisations. The speed of light in free space is 3.0×10^8 m s^{-1}. [476]

Those examples show a good method to measure ionisation energies. The shortest wavelength in the line spectrum relates to ionisation energy.

The ionisation energy of an atom is 10^{-17} J. What is the shortest wavelength in its line spectrum? [459]

A second method involves ionisation by collision. This time the radiation used is a beam of fast electrons. Again, we can use examples to show the approach.

The ionisation energy of neon is 3.4×10^{-18} J. What smallest speed must an incident electron have to cause ionisation by collision? The electron mass is 10^{-30} kg.

$W = \frac{1}{2} mv^2 \rightarrow v = \surd (2 W/m_e)$
$= \surd (2 \times 3.4 \times 10^{-18}/10^{-30})$ m s^{-1} = 2.6 Mm s^{-1}

We can find the p.d. through which the beam electrons must be accelerated to reach this energy. We use $W = VQ$ (section 18.2).

The ionisation energy of neon is 3.4×10^{-18} J. Through what smallest p.d. must an electron beam be accelerated to cause ionisation by collision? The electron charge is 1.6×10^{-19} C. [387]

This is the **ionisation potential difference** for neon.
We can measure this as follows.
A beam from an electron gun passes through the substance under test. This is in the state of a low-pressure gas. The beam energy (source voltage) is raised until ionisation takes place; the ionisation energy is found from the ionisation potential difference.

In such an experiment, sodium atoms are ionised when the electrons are accelerated through 5.1 V. What is the ionisation energy of sodium? The electron charge is 1.6×10^{-19} C. [365]

What wavelength rays would just ionise sodium? [356]

In practice, the use of spectra is far simpler than that of electron beams; it also gives better results. The electron beam method was of historical importance.

19.2 The nucleus

The nucleus is the tiny, massive, positive core of an atom.

Nuclei are around 10^{-14} m diameter – too small ever to be seen by a microscope. Yet we think we know quite a lot about nuclei; they are very complex particles.

Here we give only a simple picture. It can, however, explain some of the basic facts.

In this simple model, nuclei consist of two types of particle (or **nucleon**) – the **proton** and the **neutron**. Thus there are three 'sub-atomic particles' to know about; Table 19.2 gives the main details.

Table 19.2 The three main sub-atomic particles

Particle	Place in atom	Mass/a.m.u.	Charge/a.c.u.
Electron	cloud	0	−1
Proton	nucleus	1	1
Neutron	nucleus	1	0

The mass of a nucleon is in fact about 1.7×10^{-27} kg. This is almost 2000 times the electron mass. So it is quite in order to give the particle masses as 1 and 0 atomic mass units (a.m.u.). We already know the atomic charge unit (a.c.u.); it is 1.6×10^{-19} C.

The elements differ in the number of protons in the nucleus. This value is called the **proton number**, Z. (Some people call it the atomic number.)

The proton number of an atom is the number of protons in the nucleus.

As we know, the number of electrons in the cloud round the nucleus will tend to be the same. However, ionisation is common, so elements should not be described by the electron number.

There are no simple rules about the number of neutrons in a nucleus. Very roughly, however, the neutron number, N, lies between 1 and 1.5 times the proton number Z.

For any species of atom – the word **nuclide** is better – we define A, the **nucleon number**. (Some people call this the mass number or relative atomic mass.)

The nucleon number of an atom is the number of protons and neutrons in the nucleus.

Nucleon number, A = proton number, Z + neutron number, N

Note: A, N and Z are all numbers; they have no units.

What is the simplest atom? It is the atom with only one proton – $Z = 1$. In the neutral state, it will have one electron in the electron cloud. No neutrons are needed in this simplest atom. It is the atom of hydrogen, H.

We can describe this nuclide as in Table 19.3.

Table 19.3 The structure of the hydrogen atom

Element number	Name	Symbol	Proton number	Neutron number	Electron number	Z	A
1	Hydrogen	H	1	0	1	1	1

What is the next-simplest nuclide? It is not one with two protons – that will need two electrons and, in fact, at least one neutron. The next-simplest nuclide is a second form of hydrogen, having one neutron rather than none.

The mass of this atom – given by A – is twice that of 'normal' hydrogen – this is deuterium, D, better known as 'heavy hydrogen'. ('Heavy water' is D_2O, more than 10 per cent denser than H_2O.) Only 0.015 per cent of hydrogen is deuterium.

We can add a second neutron. This makes tritium, T – 'very heavy hydrogen'. However, like so many nuclides we could construct, tritium is not stable. It is in fact **radio-active** – the nucleus breaks up with the release of energy. In these tables we mark radio-active nuclides with an asterisk*.

Table 19.4 The isotopes of hydrogen

Element number	Name	Symbol	Proton number	Neutron number	Electron number	Z	A
1	Hydrogen	H	1	0	1	1	1
	Deuterium	D	1	1	1	1	2
	Tritium	T	1	2	1	1	3*

Table 19.4 shows the three forms of hydrogen that can exist. They have the same proton number but different nucleon numbers. The electron number depends on proton number; the electron number fixes the chemical nature of a substance. The mass depends on nucleon number; it fixes the physical nature of a substance. Thus the three **isotopes** of hydrogen have the same chemical nature; their physical natures differ.

Isotopes are different forms of an element; they have different numbers of neutrons in the nucleus. Chemically they are the same; physically they differ.

Table 19.5 The first few nuclides after hydrogen

Element	Nuclide	Proton number	Neutron number	Electron number	Z	A
2						
Helium	$^{3}_{2}$He	2	1	2	2	3
	$^{4}_{2}$He	2	2	2	2	4
	$^{6}_{2}$He	2	4	2	2	6*
3						
Lithium	$^{6}_{3}$Li	3	3	3	3	6
	$^{7}_{3}$Li	3	4	3	3	7
	$^{8}_{3}$Li	3	5	3	3	8*
	$^{9}_{3}$Li	3	6	3	3	9*

Table 19.5 gives details of the next few nuclides. You do not need to know the details, but the ideas are useful. In this table we use this common nuclide symbol – $^{A}_{Z}$ X. X is the chemical symbol of the element. Thus $^{2}_{1}$H is deuterium, with nucleon number 2 and proton number 1.

Describe the structure of $^{27}_{13}$Al. [358]

Describe the structure of $^{235}_{92}$U. [496]

19.3 The mass spectrometer

This is used to split a beam of cations (positive ions) on the basis of their mass. The output is a line spectrum with a mass axis. (The spectra of waves met before have a wavelength axis.)

In other words, the mass spectrometer splits a beam into different beams of the different isotopes.

The concept is due to J.J. Thomson. Indeed, the design of part of this device is very like his specific charge experiment described in section 18.3.

First the cation beam from a gun passes into a speed selector, Fig. 19.3(a).

The applied electric and magnetic fields, E and B_a, balance only for particles of one speed, v. Only these, therefore, pass through the slit to the spectrometer itself.

For the speed selector, then, $B_a Q_i v = E Q_i$.

Here Q_i is the ion charge.

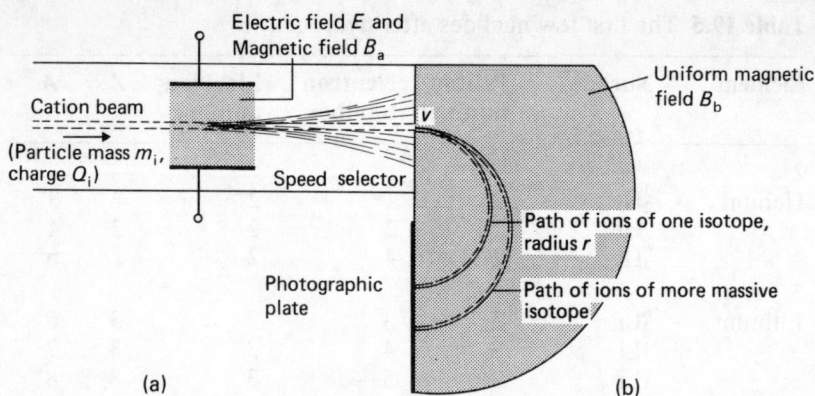

Fig. 19.3 The mass spectrometer

The speed chosen, v, is $\qquad v = E/B_a$.

We can change v simply by changing E (or B_a).

In the spectrometer itself, the ions enter a second magnetic field, B_b, at 90°. (See Fig. 19.3(b).) They are forced into a circular path. After moving through half a circle, they strike a photographic plate to cause an image.

The radius r of the path of ions of mass m_i is found as follows.

$$B_b Q_i v = m_i v^2/r \qquad (BQv \text{ is the centripetal force } mv^2/r.)$$

This relation gives the **specific charge**, Q_i/m_i:

$$Q_i/m_i = v/rB_b$$

But $\qquad v = E/B_a \qquad$ (from above)

So $\qquad Q_i/m_i = E/rB_aB_b$

For ions of the same charge, Q_i, the path radius r in a given machine depends only on the ion mass m_i (as Q_i, E, B_a and B_b are constant):

$$r \propto m_i$$

A beam of mixed $^{35}_{17}Cl$ and $^{37}_{17}Cl$ ions of the same speed enters a mass spectrometer. The radius of the path of the former is 200 mm. What is the path radius of the latter?

$m_a = 35$ a.m.u. $\quad m_b = 37$ a.m.u. $\quad r_a = 0.20$ m $\quad r_b = ?$
$r_a/r_b = m_a/m_b$ (as $r \propto m$)
$\rightarrow r_b = r_a m_b/m_a = 0.20 \times 37/35$ m $= 211$ mm

Each isotope beam follows half of a circle. In the case in the above example, therefore, images will appear on the plate 400 mm and 422 mm from the slit. When developed, the plate will look like Fig. 19.4.

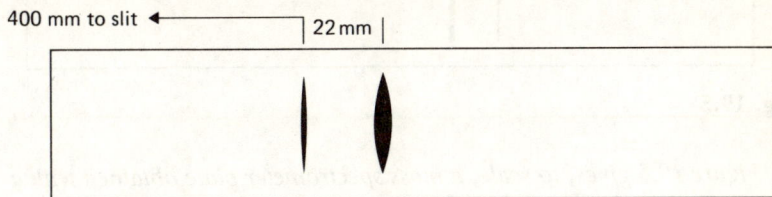

Fig. 19.4 Results of using a mass spectrometer

The second image is denser. This means that there is more of that isotope in the sample than of the other.

The three stable isotopes of magnesium are $^{24}_{12}Mg$ (80 per cent), $^{25}_{12}Mg$ (10 per cent), and $^{26}_{12}Mg$ (10 per cent). A magnesium cation beam of a certain speed enters a mass spectrometer. Three images are obtained on the film. The centre image is 300 mm from the slit. How far from this image are the other two? Describe the images. [471]

19.4 Some more questions

1. *Table 19.6 gives some ionisation energies. Complete the table by showing (i) the ionising radiation wavelength, λ; (ii) the minimum ionising electron speed, v; (iii) the ionising p.d. V. Use $Q_e = 1.6 \times 10^{-19}$ C; $m_e = 10^{-30}$ kg; $c = 3.0 \times 10^8$ m s^{-1}.*

Table 19.6

Element	Ionisation energy/J	(i) λ	(ii) v	(iii) V	
(a) Lithium Li$^+$	1.2×10^{-17}				[465]
(b) Carbon C^{++}	7.7×10^{-18}				[368]
(c) Copper Cu	1.2×10^{-18}				[447]

2. *Which one of the data which follow is wrong? (a) The ionisation energy of sodium is 8.2×10^{-18} J. (b) The ionisation electron speed for sodium is 1.3 Mm s^{-1}. (c) The ionisation p.d. for sodium is 5.1 V. Use the same constants as for Question 1.* [485]
3. *Describe the structures of arsenic nuclide $^{75}_{33}As$ [376], copper nuclide $^{65}_{29}Cu$ [398], and argon nuclide $^{40}_{18}A$ [458].*
4. *Draw your own diagram of a mass spectrometer with speed selector. Explain in your own words how it is used.* [484]

244

200 mm to slit ◄────────────

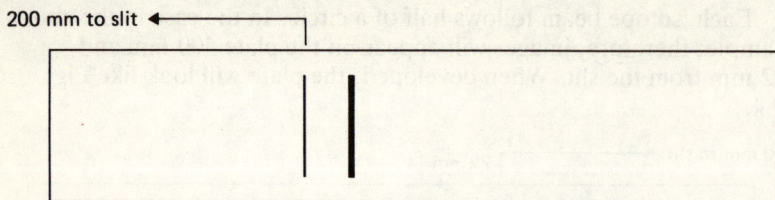

Fig. 19.5

5. *Figure 19.5 gives, to scale, a mass spectrometer plate obtained with a copper sample. The value of A for one copper isotope is 65. What can you deduce from the plate?* [467]
6. *The stable isotopes of selenium are as follows. $^{74}_{34}Se$ (1 per cent); $^{76}_{34}Se$ (10 per cent); $^{77}_{34}Se$ (7 per cent); $^{78}_{34}Se$ (23 per cent); $^{80}_{34}Se$ (50 per cent); $^{82}_{34}Se$ (9 per cent). Sketch and discuss the output you would expect from a mass spectrometer test.* [395]

19.5 Objectives

When you have studied this chapter, you should be able to

(1) define ion;
(2) define ionisation energy;
(3) explain how this can be measured from line spectra;
(4) explain how it can be measured as a result of collisions between electrons and atoms;
(5) solve simple problems on these two situations;
(6) discuss the basics of atomic and nuclear structure in terms of electrons, protons and neutrons;
(7) define and relate proton number, electron number, neutron number and nucleon number;
(8) define isotopes;
(9) explain the functions of the parts of a mass spectrometer.

Appendix I

Symbols used in this book

The main simple symbols met in this book are listed here alphabetically. Quantity symbols are *italic*; unit symbols are upright.

This list does not include symbols of more than one letter, such as Pa, pascal. Nor does it include **subscripts**. Subscripts are not standard, although many are widely used by convention. Numerical subscripts – such as in v_1, starting velocity, and v_2, final velocity – refer either to measures which change in sequence or to numbered parts of some system. Letter subscripts refer to lettered parts of some system. The subscript zero – as in V_0, peak potential difference – relates to a basic value.

1.1 Symbols using roman and italic letters

a	acceleration; amplitude
A	area; nucleon number (atomic mass number)
A	ampere (amp)
b	breadth
B	magnetic (force) field strength
c	specific thermal capacity; speed; speed of light in free space $(3.00 \times 10^8 \text{ m s}^{-1})$
C	capacitance
c	centi- (10^{-2})
C	Coulomb; Celsius
d	distance

d	differential coefficient
d	deci- (10^{-1})
e	electron charge – in this book Q_e is used for this – 1.6×10^{-19} C
e	base of natural logarithms (2.718)
E	Young's modulus; electromotive force; electric (force) field strength
F	force
F	farad
g	(local) gravitational (force) field strength; (local) acceleration of free fall (9.8 (around 10) m s^{-2})
g	gram
G	giga- (10^9)
h	Planck constant $(6.6 \times 10^{-34}$ J s)
h	hour = 3600 s
H	henry
i	angle of incidence
I	electric current
J	joule
k	force constant; used for unspecified constant
k	kilo- (10^3)
K	kelvin
l	length
L	specific latent thermal capacity; inductance
m	mass
m	metre, milli- (10^{-3})
M	mutual inductance
M	mega- (10^6)
n	unspecified number; number in sample; refractive constant (index)
n	nano- (10^{-9})
N	number of particles per unit volume; number of turns in coil; neutron number
N	newton
p	momentum; pressure
p	pico- (10^{-12})
P	power
Q	electric charge
r	radius; specific gas constant; angle of reflection; angle of refraction; source resistance
R	resistance; load resistance
s	displacement; arc length
s	second
t	time
t	tonne (= 1000 kg)
T	temperature; period; tension
T	tesla
u	unit

v	velocity; speed
V	volume; potential difference
W	weight; work; energy
x	rectangular coordinate; distance; unspecified unknown
X	reactance
X	radiation type
y	rectangular coordinate; unspecified variable
z	rectangular coordinate
Z	proton number (atomic number)

I.2 Symbols using Greek letters

α	specified angle
γ	surface tension
γ	radiation type
$\triangle$	difference between values; change of
ε	strain
η	efficiency; viscosity
θ	angular displacement; specified angle
λ	wavelength
μ	radiation type
μ	micro- (10^{-6})
ν	frequency
π	ratio of circumference to diameter of a circle (3.14159)
ρ	density
σ	stress
Σ	sum of terms like
ω	angular velocity; pulsatance (angular frequency)
Ω	ohm

Appendix II

Measures used in this book

The main quantities used in this book are set out below; the standard unit symbols (if any) are given. Also given are the defining relations if appropriate, and in the same way the sections of the book in which they are first discussed in detail.

Quantity/Unit	Relation	Section
acceleration, a/m s^{-2}	change of velocity/time	1.2
acceleration of free fall, g/ms^{-2}	–	2.2
amplitude, a and other symbols/various	–	10.1
angle, θ and other symbols/rad	arc length/radius	3.2
angular velocity, ω/rads^{-1}	angular displacement/time	3.2
area, A/m^2	–	–
capacitance, C/F	charge/potential difference	17.2
charge, Q/C	–	15.2
cosine of angle, cos/–	adjacent/hypotenuse	–
current, I/A	charge/time	14.1
density, ρ/kg m^{-3}	mass/volume	7.1
displacement, s/m	–	–
efficiency, η/–	useful energy output/input	5.3
elastic modulus – see Young's modulus		
electric current – see current		

electric field strength, E/NC^{-1} (Vm^{-1})	force/charge; p.d./distance	18.1
electromotive force, E/V	supply current × circuit resistance	14.5
energy, W/J	various	4.4
focal distance, f/m	–	10.6
force, F/N	mass × acceleration	1.3
force constant, k/Nm^{-1}	force/displacement	13.3
frequency, v/Hz	–	9.1
gravitational field strength, g/N kg^{-1}	weight/mass	2.2
inductance, L/H	–	17.3
internal resistance – see source resistance		
kinetic energy, W/J	½ × mass × velocity^2	4.4
length (distance), l/m	–	–
magnetic field strength, B/T	force/current × conductor length	15.2
mass, m/kg	–	–
momentum, p/kg m s^{-1}	mass × velocity	1.3
mutual inductance, M/H	–	16.2
neutron number, N/–	number of neutrons in atom	19.2
nucleon number, A/–	number of nucleons in atom	19.2
period, T/s	l/frequency	3.2
potential difference, V/V	energy/charge	14.1
potential energy (gravitational), W/J	mass × g × height	5.2
power, P/W	energy/time	5.3
pressure, p/Pa	force/area	7.1
proton number, Z/–	number of protons in atom	19.2
pulsatance, ω/Hz	2 π × frequency	13.2
resistance, R/Ω	potential difference/current	14.1
sine of angle, sin/–	opposite/hypotenuse	–
source resistance, r/–	Ω	14.5
specific gas constant, r J kg^{-v} K^{-1}	(pressure × volume)/ (mass × temperature)	7.7
specific latent thermal capacity, L/J kg^{-1}	energy/mass	9.2
specific thermal capacity, c J kg^{-1} K^{-1}	energy/(mass × temperature change)	9.2
speed, c/m s^{-1}	distance/time	9.1
strain, ϵ/–	length change/length	6.5
stress, σ/N m^{-2}	force/area	6.4
surface tension, γ/J m^{-2}	energy/area	9.3
temperature, T/K	–	7.3
time, t/s	–	–
velocity, v/m s^{-1}	displacement/time	1.2

viscosity, η/Pas	–	9.4
volume, V/m^3	–	10.1
wavelength, λ/m	–	10.1
weight, W/N	mass $\times g$	1.3
work, W/J	force $\times$ distance	4.4
Young's modulus, E/N m^{-2}	stress/strain	6.6

Appendix III

Answers to questions

All the unworked questions in this book have been given a code like [123] in random order. The answers, hints and comments for each question appear here in numerical order of code number.

100. speed = distance/time = 1500 m/212 s = 7.07 m s^{-1}.

101. 10.8 m s^{-1}, 7 °N or S of E (depending which way the sailor runs).

102. 600 N – approximately the weight of a woman.

103. The tension in the sling. The sling is held by the user's hand – the user pulls the sling and the stone all the time.

104. (a) 1.8 s; (b) 12.5 N. In this arrangement (the 'conical pendulum'), the load's weight equals the vertical component of the tension; the horizontal component is the centripetal force.

105. (a) 106 m s^{-2}; (b) 25 kN; (c) 12.5 kJ (from $W = Fs$).

106. (a) 15 000 kg m s^{-1}; (b) 22.5 kJ.

107. 40 J.   **108.** 60 J.   **109.** 650 N (the answer must include the force to overcome friction).

110. (a) 5.2 m s^{-1}, 58 J; (b) 1.2 m s^{-1}, 314 J. The total kinetic energy after impact is less than before – but there is no 'loss': energy is transferred to sound, temperature rise, deformation, etc.

111. $m = 0.005$ kg $v = 0.5$ m s^{-1} $p = ?$
$p = mv = 0.005 \times 0.5$ kg m s^{-1} = 0.0025 kg m s^{-1}.

112. 24 °C (relating $W = Pt$ and $W = mc\Delta T$)

113. (b) 110J: $W = (10 \times 2) + (15 \times 2) + (20 \times 2) + (10 \times 2)$ J.

114. 1 C, 2T, 3T, 4C. (C-compression, T-tension).

115. $V_1 = 10^{-2}\,m^3$ $p_1 = 5 \times 10^5\,Pa$ $V_2 = 10^{-3}\,m^3$ $p_2 = ?$
$p_1V_1 = p_2V_2 \rightarrow p_2 = 5 \times 10^5 \times 10^{-2}/10^{-3}\,Pa = 5\,MPa.$

116. 450 K (177 °C).

117. 5 000 000 000 kg m s^{-1} (5×10^9 kg m s^{-1}).

118. 8.5 m s^{-1}. **119.** 750 K (477 °C). **120.** The total mass
is the sum of the masses – 1100 kg.

121. 2.6 MPa. **122.** $T = 273\,K$ $p = 10^5\,Pa$ $\rho = 0.09\,kg\,m^{-3}$
$\sqrt{(v^2)} = ?$

$$pV = \tfrac{1}{3} Nm_p \overline{v^2} \rightarrow p = \tfrac{1}{3} \rho\overline{v^2} \rightarrow v^2 = 3p/\rho.$$

$$\rightarrow \sqrt{(v^2)} = \sqrt{(3p/\rho)} = \sqrt{(3 \times 10^5/0.09)}\,m\,s^{-1} = 1.8\,km\,s^{-1}.$$

123. The size of the box makes no difference in the end.

124. The short answer is 'yes'! Energy is needed to cause boiling.
When boiling occurs, more energy must be supplied. If not, the
temperature will fall, and boiling will cease.

125. 37 °C.

126. 2 km.

127. 470 m s^{-1}.

128. (a) 19.8 kJ; (b) 16.3 kJ.

129. 50 kHz.

130. $F = 8 \times 10^{-12}\,N$ $a = 4 \times 10^{-3}\,m\,s^{-2}$ $m = ?$
$m = F/a = 2 \times 10^{-9}\,kg = 2\,\mu g$ ($2 \times 10^{-6}\,g$).

131. $g = 1.5\,m\,s^{-2}$ $r = 1.8 \times 10^6\,m$ $v = ?$

$F = mv^2/r \rightarrow mg = mv^2/r \rightarrow v = \sqrt{(gr)} = \sqrt{(1.5 \times 1.8 \times 10^6)}\,m\,s^{-}$
$1.64 \times 10^3\,m\,s^{-1}.$

Note that this answer does not depend on the mass of the
spacecraft. Why not?

132. $i = 0°$ $r = ?$
$r = i = 0°$. The ray comes back along its original path.

133. Northwest; 112½° anticlockwise.

134. Compare Fig. 11.2.

135. (a) 1.2 kg m s^{-1}; (b) 0.86 kg m s^{-1}.

136. Your essay should confirm that if a wave carries energy, it can
do work. The energy in the wave can move something, or it can
appear as a second type of energy. Examples have been given in
section 10.1. There are many others.

137. 200 kHz; 1.32×10^{-28} J; radio.

138. 1.65 m. **139.** 2.0 m. **140.** (Use the same method as
in the previous question, 240) 6.75 m.

141. (a) 125 mm beyond lens, 2.5 mm, real, inverted;
(b) 33 mm on object side of lens, 13.3 mm, virtual, upright.

142. 0.1 A.
143. 2590 m s^{-2}.
144. 3.9 × 10^9 Nm^{-1}.
145. 50 Ω.
146. 1.1 Hz.
147. 50 m s^{-1}.
148. The distance between fringes will be halved.
149. You should bring in Fig. 12.1 and 12.3 and discuss how the light is spread into a spectrum in each case.
150. $m = 0.05$ kg $v = 1000$ m s^{-1} $p = ?$
    $p = mv = 0.05 × 1000$ kg m s^{-1} $= 50$ kg m s^{-1}.
151. 80 m – equate mgh with $\frac{1}{2}mv_2{}^2$. See next few paragraphs of text.
152. (a) 40 m s^{-1}; (b) 60 m.
153. See Fig. 10.4 and accompanying text.
154. 5 × 10^{15} Hz; 3.3 × 10^{-18} J; ultra-violet.
155. 7500 N (= 15 000 × sin 30°N).
156. Compare with Fig. 11.2 and the text with it.
157. 20 A.
158. 40 Hz.
159. (a) 10 m s^{-1}, (b) 17 m s^{-1}.
160. 6 kN.
161. $p_1 = 50 × 10^5$ Pa $T_1 = 300$ K $T_2 = 900$ K $p_2 = ?$

    $p_1/T_1 = p_2/T_2 \rightarrow p_2 = p_1\ T_2/T_1$
    $= 50 × 10^5 × 900/300$ Pa $= 15$ MPa.

162. On the Moon g is 1.5 m s^{-2} (given a few lines earlier); again $v_1 = 0$ m s^{-1}. Answer 4.5 m s^{-1}.
163. 10^{-12} m; 1.98 × 10^{-13} J; gamma.
164. (a) 1 μC; (b) 12.5 C.
165. $R_a = R_b$. The display is powered by either V_a or V_b.
166. 333 mm.   167. 2.7 MV.
168. The value of V rises steeply from zero at fist, but then more and more slowly. It will reach 150 V (= e.m.f.) when the current reaches zero. Compare with the definition of electromotive force.
169. 10 Ω.   170. 27 MPa.
171. (a) 100 N; (b) 173 N.
172. 5 μm.
173. 100 Ω, 100 Ω.
174. 1.5A (assuming the meter resistance is zero).
175. The roller moves right (from the left-hand motor rule).
176. 0.105 V.

177. The wires will be forced together in the first case and apart in the second. The magnetic field patterns should show this if field lines are pictured as elastic. A useful law of physics that is not well known is – 'Like currents attract each other; unlike currents repel each other.'

178. (c) 20 kV; (d) 5 mV.

179. 7.9×10^{14} Hz; 3.9×10^{14} Hz.

180. Answers: 125 kN, 188°. See also comment 215, for previous question.

181. $m = 100\,000$ kg $\quad a = 50$ m s^{-2} $F = ?$
$F = ma = 100\,000 \times 50$ N $= 5$ MN (or 5 000 000 N) (not unusual for a spacecraft launch).

182. $m = 10^{-6}$ kg (*not* 10^{-3} kg) $\quad a = 1000$ m s^{-2} (or 10^3 m s^{-2}) $\quad F = ?$
$F = m\,a = 10^{-6} \times 1000$ N $= 10^{-3}$ N.

183. 2400 V; 62.5 A. **184.** 3.2×10^{12} J.

185. (a) Energy turns magnet; charge is forced round coil; magnetic field set up in meter; needle turns against spring.
(b) Energy turns electret; magnetic field set up in the space around; needle turns against Earth's field.
Note: the question said 'discuss with care'!

186. There are many useful examples in this book. Your house contains dozens more!

187. 10^{-2} m; 1.98×10^{-23} J; microwave

188. Infinity (note that the frequency of d.c. is zero – and capacitors block d.c.).

189. The force on the needle will reverse many times a second. Thus the needle will only vibrate at the zero position. Standard moving-coil meters cannot record a.c.; others can.

190. *Zero* – in any cycle, the negative section just equals the positive section.

191. Both have a momentum of 250 kg m s^{-1}.

192. 20 V.

193. The centre of the pattern will be a white fringe (all wavelengths interfere constructively there). On each side will be a confused multicoloured pattern (*not* a colour spectrum) – the fringes for each wavelength do not lie in the same places.

194. $E_2 = 1000$V $\quad E_1 = 50$ V $\quad N_2/N_1 = ?$

$N_2/N_1 = E_2/E_1 = 1000/50 = 20.$

195. 912 W. **196.** 2.1 T.

197. 160 Ω; $160\mu\Omega$.

198. 53 V.

199. Again we can use $F = m\,(v_2 - v_1)/t \rightarrow$
$t = 10^5 \times (6 \times 10^3 - 0)/2 \times 10^6 = 300$ s

200. Initially the ball is at rest – $v_1 = 0$ ms^{-1} $\quad v_2 = 10$ ms^{-1} $\quad t = 0.25$ s.
Answer 40 m s^{-2}.

201. 0.5 m.   **202.** Using $m = 5 \times 10^{-3}$ kg and $a = 50$ m s^{-2} with $F = ma$ gives $F = 250$ mN.

203. 1.98×10^{-25} m; 1.52×10^{33} Hz; gamma.

204. $\lambda = 1500$ m $c = 3 \times 10^8$ m s^{-1} $v = ?$ $T = ?$

$c = v\lambda \rightarrow v = c/\lambda = 3 \times 10^8/1500$ Hz $= 200\,000$ Hz $= 200$ kHz
$T = 1/v = 1/(2 \times 10^5)$ s $= 5\ \mu$s.

205. $m = 0.01$ kg $v = 5 \times 10^3$ Hz $k = ?$
$v = 1/2\,\pi\,\sqrt{(k/m)} \rightarrow k = 4\pi^2\,v^2 m = 4\pi^2 \times (5 \times 10^3)^2 \times 0.01$ Nm^{-1}
$= 9.9$ MNm^{-1}.

206. (b) $E_2 = 12.5$ V.
$E_1 = 250$ V $r_2 = 2.0\ \Omega$ $R = 0\ \Omega$ $I_2 = ?$ $I_1 = ?$
$I_2 = E_2/(R + r_2) = 12.5/(0 + 2.0)$A $= 6.25$ A
$I_1 = EI_2/E_1 = 12.5 \times 6.256/250$ A $= = 313$ mA.

207. (a) 1.26 MΩ; (b) 1.26 kΩ; (c) zero.

208. 1.06 V.

209. 170 V; 17 V; 340 V.

210. Attempt to provide a perfect magnetic circuit to link the conductors, each being wound into a large coil.

211. 4×10^{16} m s^{-2}.

212. (e) 200 μF; (f) 0.4 μF.

213. $F = 2.0$ N $s = 10^{-2}$ m $k = ?$
$k = F/s = 2/10^{-2}$ Nm$^{-1} = 200$ Nm^{-1}.

214. 3.362×10^{19} J, 3.358×10^{-19} J; 4.95×10^{-4}m; infra-red or microwave.

215. By scale drawing as in the previous examples:
(a) 5.8 m s^{-1}, 007°; If your answers are close but not quite the same, your drawing was not accurate enough – perhaps too small, or drawn with a blunt pencil.

216. 20 kΩ.

217. (a) 4 m s^{-2}; (b) 360 J; (c) 12 m s^{-1}.

218. $m = 40$ kg $v = 1$ m s^{-1} $p = ?$

$p = mv = 40 \times 1$ kg m s$^{-1} = 40$ kg m s^{-1}.

219. This action is like that of the rocket. The elastic balloon pushes the air out; the air pushes the balloon the other way. Newton's third law.

220. (i) 265 K (−8 °C) (ii) 388 K (115 °C) (Answers are approximate.)

221. A rough ray diagram (or set of these) should give the answer(s). The image will be somewhere on the eye side of the lens. If the image is further from the lens than the eye, nothing will be seen in focus.

222. $m = 10\,000$ kg $v = 25$ m s^{-1} $p = ?$

$p = mv = 10\,000 \times 25$ kg m s$^{-1} = 250\,000$ kg m s^{-1}

223. 4.9 A.

224. 0.71 V.

225. $F = 50$ N $m = 0.05$ kg $a = ?$

$F = ma \rightarrow a = F/m = 50/0.05$ m s^{-2} = 1 km s^{-2}.

226. 10^{-18} s; 3 $\times 10^{10}$ m.

227. Temperature is scalar – so we add – answer: 30 °C.

228. doubles.

229. $X_L = 10^3$ Ω $v = 10^3$ Hz $L = ?$

$X_L = 2 \pi v L \rightarrow L = X_L/(2 \pi v) = 10^3/(2\pi \times 10^3)$ H
$= 0.16$ H.

230. 580 N, 17° from the forward direction.

231. (a) Field lines between n- and s-poles but not between
s-poles – empty space ('neutral point') somewhere between
s-poles.
(b) Field like that of a bar magnet but inside the coil as well.
Near end of coil acts like s-pole; far end acts like n-pole.
Note: All the space shown should carry lines. Direction of fields
must be shown.

232. 1.25 N.   **233.** 0.25 A.

234. (a) 6 m s^{-1} (b) -30 kg m s^{-1}, 30 kg m s^{-1} (c) 135 J.

235. (Refer to Fig. 1.9.) $v_e = 5 \sin 22\frac{1}{2}°$ m s^{-1} = 1.9 m s^{-1}
$v_n = 5 \cos 22\frac{1}{2}°$ m s^{-1} = 4.6 m s^{-1}.

236. The double throw switch must let you (a) allow the supply
to charge the capacitor, then (b) allow the capacitor to
discharge. The ammeter and resistor are both in series with the
capacitor; the voltmeter is in parallel with it. Check with care
that your circuit fits this description (or is as good).

237. 1.98×10^{-5} m; 1.52×10^{13} Hz; infra-red.

238. 1.89 MJ.

239. 500 MPa.

240. $v_1 = 0$ m s^{-1} $t = 3$ s $a = g = 10$ m s^{-2} $s = ?$ First find v_2.

$v_2 = v_1 + at = 0 + 10 \times 3$ m s^{-1} = 30 m s^{-1}
$s = \frac{1}{2} \times (v_1 + v_2) t = \frac{1}{2} (0 + 30) \times 3$ m = 45 m.

241. 10^{11}J.

242. Your essay should include much of the early part of the chapter
in your own words. Try to do questions like this without any
help, after as much preparation as you need.

243. $m = 10\ 000$ kg $g = 10$ N kg$^{-1} \rightarrow F = W = 10^5$ N
$A = 2.5 \times 10^{-2}$ m^2 $p = ?$

$p = F/A = 10^5/2.5 \times 10^{-2}$ Pa = 4 MPa.

244. You could describe the alcohol in glass thermometer, the
electric resistance thermometer, the liquid crystal type, the

pyrometer, the gas thermometer, or the bimetallic strip type. The properties in question are, respectively, the length of the alcohol column, the resistance of the sample, the reflectivity of the substance, the sample's colour, the gas pressure, the expansions of the two metals.

245. The field strength, the charge, the mass.

246. 80 Ω.

247. 24.4 m s^{-1} in the direction of the bullet.

248. Steel 0.167 s; water 0.67 s; air 3.03 s.

249. Using a = F/m with $F = 10^6$ N and $m = 20\,000$ kg gives $a = 50$ m s^{-2}.

250. Speed = distance/time = 9000/9000 m s^{-1} = 1 m s^{-1}.

251. 4 m s^{-2}.

252. 1.4 m s^{-1} (This is $\sqrt{2}$ times the first value.)

253. Both measures fall in step (to zero at the critical temperature). This follows the behaviour of the particles as temperature rises. (See section 9.3.)

254. $P = 5000$ W $E = 250$ V $I = ?$

$$P = EI \rightarrow I = P/E = 5000/250 \text{ A} = 20 \text{ A}.$$

255. 0.8 mΩ.

256. (A circuit diagram would help.)

$$R_1 = 20\ \Omega\ R_2 = 60\ \Omega\ R_p = ?$$

$$1/R_p = 1/R_1 + 1/R_2 = 1/20 + 1/60\ \Omega^{-1} = 4/60\ \Omega^{-1} \rightarrow R_p = 15\ \Omega$$

$$V = 90\ V\ I = 1\ A\ R = ?$$

$$R = V/I = 90/1\ \Omega = 90\ \Omega$$

$$R = 90\ \Omega\ R_p = 15\ \Omega\ R_s = ?$$

$$R = R_p + R_s \rightarrow R_s = R - R_p = 90 - 15\ \Omega = 75\ \Omega.$$

257. 0.3 T.　　**258.** 1.4 V.

259. The same in all cases (as long as the length units are the same).

260. (a) 5 m s^{-2}; (b) 67.5 m.

261. 4.95 km s^{-1} (the velocity of the probe is 6 km s^{-1}).

262. Think of a reflection grating as a transmission grating backed by a mirror.

263. $F = 10$ N $Q = 10^{-4}$ C $B = 100$ T $v = ?$

$$F = BQv \rightarrow v = F/BQ = 10/(100 \times 10^{-4})\text{m s}^{-1} = 1 \text{ km s}^{-1}.$$

264. 110 V.

265. 0.55 Ω, 0.83 Ω, 1.3 Ω, 1.5 Ω, 2.2 Ω, 2.8 Ω, 3.7 Ω, 6.0 Ω.

266. $m = 4$ kg $a = g = 10$ m s^{-2} $F = W = ?$

$$F = ma \text{ (or } W = mg) = 4 \times 10 \text{ N} = 40 \text{ N (the weight of four bags of sugar).}$$

267. 48.

268. 2×10^7 J (using $W = Fs$ or $W = \frac{1}{2}mv_2{}^2$).

270. 6 °C (6 K) (relating $W = VIt$ and $W = m c \triangle T$).

271. $\lambda = 3 \times 10^{-6}$ m $c = 3 \times 10^8$ m s^{-1} $v = ?$

$c = v\lambda \rightarrow v = c/\lambda = 3 \times 10^8/3 \times 10^{-6}$ Hz $= 10^{14}$ Hz.

272. $C = 50 \times 10^{-12}$ F $Q = 10^{-7}$ C $V = ?$

$V = Q/C = 10^{-7}/(50 \times 10^{-12})$ V $= 2$ kV.

273. 90 mN.

274. 100 kN (using $F = ma$).

275. $v_1 = 20$ m s^{-1} $v_2 = 0$ m s^{-1} $t = 2$s $s = ?$

$s = \frac{1}{2}(v_1 + v_2) t = \frac{1}{2} \times (20 + 0) \times 2$ m $= 20$ m.

276. Your answer should state the law and show that it would apply to a working alternator (compare with the electric motor). The result is that the alternator coil is harder to turn than if the magnets weren't there

277. 12 V, 2 Ω.   **278.** 0.1 m.   **279.** 125 rads^{-1}.

280. 59 minutes.

281. (i) 692.6 K; (ii) −218.6 °C; (iii) 20 K; (iv) −259 °C.

282. Compare with the moving coil meter in such a situation – answer 189. This time, however, the device is *supposed* to vibrate around its position!

283. 4/3 units.

284. $R = 5 \times 10^3$ Ω $I = 20 \times 10^{-3}$ A $V = ?$

$V = IR = 20 \times 10^{-3} \times 5 \times 10^3$ V $= 100$ V.

285. 7 m s^{-1}.

286. The exact answers are −86.5 °C, −78.3 °C.

287. 100 GJ.   **288.** 53 μV.

289. The Earth's gravity – the Earth is at the centre of the orbit in this case.

290. 8 km (using $p = h\rho g$).

291. 15 N.

292. 12/7 A.

293. 1 T, 2 T, 3 C, 4 T, 5 C, 6 C.

294. (a) 910 N; (b) 420 N.

295. 84 per cent.

296. The values of E/MNm^{-2} are, in order, 600, 30, 10. The differences arise from the fibrous structure of wood and its method of growth. The idea of 'grain' in woodwork is relevant.

297. Although the graph is drawn accurately for the range concerned, it is not too easy to use. The *actual* values asked for follow. Your answers may not all be very close. Don't worry about this. It would be interesting to compare your answers with those of other people.

(a) (i) 3.6 kPa (ii) 234 kPa; (b) (i) 82 °C (ii) 112 °C;
(c) (i) 12.3 kPa (ii) 1550 kPa; (d) 374 °C.
298. 34.1 m s^{-1} in the direction of the bullet.
299. (a) 211 m s^{-1}; (b) 453 m s^{-1}.
300. (a) 31.5 Ms; (b) 0.20 μrad s^{-1}; (c) 30 km s^{-1}; (d) 6.0 mm s^{-2};
(e) 3.6 × 10^{22} N; (f) 6.0 mm s^{-2}.
301. 2.5 kJ.　　**302.** 10 900 m^3.
303. 167 A, 12.1 A.
304. (a) 450 m s^{-1} for both plane and hailstone (ignoring air
friction).
(b) One point of view follows from $v_2 = v_1 + gt$. Another
would use $\frac{1}{2}mv_2^2 = mgh$.
305. 3.2GNm^{-2}.
306. 0.04 V. (*Note*: R_a and R_b are in series, so $V_1 = V_a + V_b$.)
307. 6.3 Ω; 6.3 MΩ.
308. 1350 km.
309. $r = v/BQ$.
310. (a) This is the scalar distance – 22 m; (b) Here we have the
vector displacement – 8 m North.
311. 20 J.
312. $p_1 = 4 \times 10^7$ Pa $p_2 = ?$ $T_1 = 300$ K $T_2 = 100$ K $V_1 = V_2$

$p_1V_1/T_1 = p_2V_2$ $T_2 \rightarrow p_2 = p_1T_2/T_1$
$= 4 \times 10^7 \times 100/300$ Pa $= 1.3 \times 10^7$ Pa.
313. 6 K (6 °C). This follows finding the energy removed by
evaporation. This is $W = mL = 5$ MJ. Use of $W = mc \triangle T$ for
the water that remains gives $\triangle T = 6$ K.
314. (a) 100 kN (as before); (b) 8 × 10^7 J (energy 'lost' to 'heat',
noise, friction, etc.).
315. Speed = distance/time = 5 m/1 s or 25 m/5 s (for
instance) = 5 m s^{-1}
316. $v = 10^{16}$ Hz $c = 1.5 \times 10^8$ m s^{-1} $\lambda = ?$

$c = v\lambda \rightarrow \lambda = c/v = 1.5 \times 10^8/10^{16}$ m $= 1.5 \times 10^{-3}$ m $= 15$ nm.

317. $F \alpha \sin \theta$.
318. $I_2 = 500 \times 20$ A $E_2 = 250$ V $P_2 = ?$ $\eta = 0.95$ $P_1 = ?$

$P_2 = E_2I_2 = 250 \times 500 \times 20$ W $= 2\ 500\ 000$ W $= 2.5$ MW

$\eta = P_2/P_1 \rightarrow P_1 = P_2/\eta = 2.5 \times 10^6/0.95$ W $= 2.63$ MW.
319. 5 Ω.
320. There are three stages here, best shown by a sketch graph. The
total work done, W, is the sum of the work done in each stage,
$W_1 + W_2 + W_3$. By graph and use of $W = Fs$, we have
$W = (5 \times 2) + (10 \times 2) + (20 \times 5)$ J $= 130$ J.
321. 10^7 Nm^{-2} (1 m^2 = 10^6 mm^2).
322. The component of the velocity of the wires concerned through
the field is proportional to sin θ, θ being the angle of the coil

from its position in the third sketch. As $E \propto v$ (second law of electromagnetic induction), $E \propto \sin \theta$. This leads to the answer you need.

323. 173 m s^{-1}.

324. The main point to discuss is that in boiling, the particles move far further apart than in melting. This requires much energy.

325. $v_1 = 0$ m s^{-1}
$a = g = 10$ m s^{-2}
$t = 3$ s $v_2 = ?$
$a = (v_2 - v_1)/t \rightarrow v_2 = v_1 + at = 0 + (10 \times 3)$ m s^{-1} = 30 m s^{-1}
(straight down).

326. 250C.

327. Work (= energy transfer, level 1) is a scalar. Answer: 1300 J.

328. The lenses that are thinner in the centre than at the edge – (c) and (f).

329. $m = 1000$ kg $a = 2$ m s^{-2} $F = ?$

$F = ma = 1000 \times 2$ N = 2 kN (or 2000 N).

330. $l = 1.5$ m $\triangle\, l = 0.01$ m. So ϵ is 0.0067.

331. Follow through the text by Fig. 17.2 using a specific example. Then try to deal with the question in general.

332. (i) 1337 K; (ii) 962 °C.

333. 6.8 m s^{-1} S, 2.3 $\times$ 10^4 kg m s^{-1} S.

334. 300 kPa.

335. First find v_2 using $v_1 = 0$ m s^{-1} $t = 2$ s and $g = 10$ m s^{-2}. Using $v_2 = v_1 + gt$ gives $v_2 = 20$ m s^{-1}. Now $p = mv$ leads to the answer – 40 kg m s^{-1}.

336. 5730 s (1.6 h).

337. 2.7 m s^{-1} N, 2.7 $\times$ 10^4 kg m s^{-1} N.

338. Your essay should give, in your own words, the ideas on this in section 10.3.

339. $Q_e = 1.6 \times 10^{-19}$ C $Q_e/m_e = 1.75 \times 10^{11}$ C kg^{-1}
So $m_e = 9.1 \times 10^{-31}$ kg.

340. (a) 200 kN (using $F = pA$). (b) Two (if full-grown).

341. Include these points. The space in the box is always saturated. When the air pressure is reduced to the s.v.p. of water, the water will boil. (The pressure concerned is about 10 kPa, in fact.) These facts should be explained by the behaviour of the water particles.

342. This is very like the case shown in Fig. 10.9.

343. $F = 35 \times 10^{-6}$ N $m = 7 \times 10^{-6}$ kg $a = F/m = 5$ m s^{-2}

344. $E_s = 0.85$ V $l_s = 0.17$ m $E = 3.3$ V $l = ?$

$E/E_s = l/l_s$
$\rightarrow l = l_s\,(E/E_s) = 0.17 \times 3.3/0.85$ m = 0.66 m.

345. Your diagrams should be rather like Figs. 14.3 and 14.4 – but

much simpler. Make sure that you use standard symbols in all circuit diagrams.

346. 6.7×10^8 Nm^{-2}.

347. 3.70×10^{-20} J.

348. In section 8.2 we saw that a liquid boils when its s.v.p. equals the outside pressure. Thus the boiling temperature of a liquid depends very much on the pressure. The melting temperature of a solid also depends on pressure. Therefore, pressure must be defined when we define change of state temperatures.

349. $F = 5 \times 10^3$ N $a = 10$ m s^{-2} $m = F/a = 500$ kg.

350. 30 m s^{-2}.

351. $v_1 = 25$ m s^{-1} $v_2 = -25$ m s^{-1} $t = 0.25$ s $m = 0.1$ kg $F = ?$

$F = ma = m(v_2 - v_1)/t = 0.1 \times (-25-25)/0.2$ N $= -25$ N.

352. 50 Ω.

353. 22 mm.   **354.** 450 m s^{-1}.

355. The shape of the field in Fig. 18.2(d) results from the effect of negative charge appearing at the left of the sample, and positive at the other. (This effect is the electric **polarisation** of matter.) You can extend this idea to the picking up of a scrap of paper by a charged object.

356. 240 nm.   **357.** 250 kV m^{-1}.

358. This aluminium ('Al') nuclide has $Z = 13$ meaning 13 nuclear protons; it has $A = 27$, meaning 27 nucleons. This form of aluminium has 13 protons and 14 neutrons in the nucleus. If neutral, it will have 13 electrons in the electron cloud.

359. (a) 1 N; (b) 100 mm.

360. $v_1 = 0$ m s^{-1} $v_2 = 2 \times 10^7$ m s^{-1} $F = 10^{-14}$ N
$m = 10^{-30}$ kg $t = ?$
First find a, with $a = F/m$, giving 10^{16} m s^{-2}. Then use $v_2 = v_1 + at$.
Answer 2 ms.
Compare Fig. 11.2 and the previous question. This time, however, the answer is found by taking (b) away from (a) at each point.

362. $d = 0.1$ m $D = 0.25$ m $y = 0.075$ m $\lambda = ?$
$c = 3 \times 10^8$ m s^{-1} $v = ?$

$\lambda = yd/D = 0.075 \times 0.1/0.25$ m $= 30$ mm.
$v = c/\lambda = 3 \times 10^8/30 \times 10^{-3}$ Hz $= 10^{10}$ Hz.

363. 15.25 kW; 98.4 per cent.   **364.** 6.2 Mm s^{-1}.

365. $V = 5.1$ V $Q_e = 1.6 \times 10^{-19}$ C $W = ?$

$W = VQ = 5.1 \times 1.6 \times 10^{-19}$ J $= 8.2 \times 10^{-19}$ J.

366. $I_1 = ?$ $E_1 = 11\,000$ V $P_1 = 2.63 \times 10^6$ W

$P_1 = E_1 I_1 \rightarrow I_1 = P_1/E_1 = 2.63 \times 10^6/(11 \times 10^3)$ A $= 239$ A.

367. $X_c = 1.0$ Ω $v = 10^3$ Hz $C = ?$

$X_c = 1/(2 \pi v C) \rightarrow C = 1/(2 \pi v X_c) = 1/(2 \pi \times 10^3 \times 1)$ F
$= 160 \ \mu$F.

368. 26 nm, 3.9 Mm s^{-1}, 48 V.   **369.** 4.2 rads^{-1}.

370. 8×10^8 kg m s^{-1} **371.** See Fig. 12.8.

372. The set of experiments in section 16.1 are very much the same; they used a permanent magnet, while you are using an electromagnet.

373. $E_{rms} = 12$ V $E_o = ?$

$E_o = \sqrt{2}E_{rms} = 1.414 \times 12$ V $= 17$ V.

374. 25 mC.

375. The ball will 'tick' back and forth between the plates. It picks up the charge from one on contact, and is forced to the other, where it becomes charged the other way. The process will go on as long as there is charge on the plates.

376. 33 protons; 42 neutrons; 33 electrons (if the arsenic atom is neutral).

377. The equation at the end of section 18.2 in fact follows from the equations of motion. This is because we derive $W = \frac{1}{2}mv^2$ from them. (See section 4.4.)

378. Your answers may be as close as you like to section 12.5 – as long as you understand what you are doing!

379. 4/3 s.

380. (a) 500 m; (b) 3.2 m.   **381.** 5 000 000; 0.3 m s.

382. 0.99 MNm^{-1}.

383. 0.35 V.

384. $m = 0.50$ kg $\quad k = 200$ Nm^{-1} $\quad v = ?$

$v = 1/2\pi \ \sqrt{(k/m)} = 1/2\pi \ \sqrt{(200/0.5)}$ Hz $= 3.18$ Hz.

385. $I = 13$ A $V = 260$ V $R = ?$

$R = V/I = 260/13 \Omega = 20 \ \Omega$.

386. (i) 3C, (ii) 180 C.

387. $W = 3.4 \times 10^{-18}$ J $Q_e = 1.6 \times 10^{-19}$ C $V = ?$

$V = W/Q = 3.4 \times 10^{-18}/1.6 \times 10^{-19}$ V $= 21$ V.

388. 5 Hz; 15 Hz; 25 Hz.

389. 9.5 m s^{-1} S, 2.0×10^4 kg m s^{-1} S.

390. If you cannot do this question, try again after you have studied Chapter 17.

391. $m = 10$ kg $\quad a = 0.1$ m s^{-2} $\quad F = ?$

$F = ma = 10 \times 0.1$ N $= 1$ N.

392. $V_1 = 200$ m^3 $p_1 = 10^5$ Pa $V_2 = 0.5$ m^3 $p_2 = ?$ $T_1 = T_2$

$p_1V_1/T_1 = p_2V_2/T_2$
$\rightarrow p_2 = p_1V_1/V_2 = 10^5 \times 200/0.5$ Pa $= 4 \times 10^7$ Pa.

393. The answer comes best by drawing – it is 60°, whichever way the ray came, whichever way the mirror turned. The reflected ray is always turned through twice the angle turned by the mirror. This is the **optical lever** effect.

394. 20 mN, 20 N, 20 kN.

395. Develop your answer on the lines of the text with Fig. 19.5; compare the previous question (Answer 467).

396. See text following and Fig. 1.4.

397. We ask you to show you understand Thomson's experiment – by writing about it in your own words with your own sketch. The essence of it is the balance between forces.

398. 29 protons, 36 neutrons (and 29 electrons in the cloud of the neutral atom).

399. (a) 62.5 m; (b) 170 kJ.   **400.** 3.5 kJ.

401. $\lambda = 500 \times 10^{-9}$ m $c = 3.00 \times 10^8$ m s^{-1} $h = 6.6 \times 10^{-34}$ J s
$W = ?$

$$W = h\nu \quad c = \nu \lambda$$

$$\rightarrow W = hc/\lambda$$
$$= 6.6 \times 10^{-34} \times 3.00 \times 10^8/500 \times 10^{-9} \text{ J}$$
$$= 3.96 \times 10^{-19} \text{ J}.$$

402. 1 Ω.

403. The approach described only brings in resistance. The voltage transformer is a machine used with a.c., with a high inductance. Self and mutual inductance must be treated in practice. The text (section 16.4) in effect involves the latter by showing that I_1 depends on the resistance of the output circuit.

404. 0.5 mV.

405. 0.8 Ω.

406. See Answer 397 – the same applies.

407. See following text.

408. Fluorescence is the absorption of high-energy quanta, leading to the emission of lower-energy photons. Excitation gives 'falling-down' by two or more stages.

409. Here we use $m = F/a = 4$ kg.

410. 73 μrads^{-1}.

411. (a) 150 kg m s^{-1}; (b) 0.5 m s^{-1}. Both are the opposite way to the jump.

412. (a) 200 kN; (b) 4×10^7 Nm^{-2}.

413. 2.3 mA (this relates to a temperature of 37 °C).

414. Above the critical temperature the liquid cannot exist whatever the pressure. Thus at the critical temperature no energy is used to boil the liquid; there is no surface tension to overcome.

415. Your essay should give, in your own words, the ideas on this in section 10.2.

416. A 10 Ω variable resistor is in parallel with a series of three 10 Ω

264

resistors. The combination is supplied with 12 V d.c., the source resistance being 2 Ω, there being an ammeter, a switch and a 3 A fuse.

417. Total pressure is 50×100 kPa $= 5 \times 10^6$ Pa $= 5 \times 10^6$ Nm^{-2}. From the previous question [192], each impact gives a force on the wall of 1.4×10^{-13} N. *If each impact is at 90°*, there are $5 \times 10^6/1.4 \times 10^{-13}$ impacts per second per unit area $= 3.6 \times 10^{19}$. This is an *average* figure.

418. 5.35 μm.

419. 0.75 N.

420. $m = 0.5$ kg $a = 30$ m s^{-2} $F = ?$

$F = ma = 0.5 \times 30$ N $= 15$ N.

421. 37 J kg^{-1} K^{-1}.

422. 10 m (using $W_1 = Pt$, $W_2 = \eta W_1$, $W_2 = mgh$).

423. The pressure is halved.

424. 346 m s^{-1}.

425. Note the word 'explain'!

426. 330 m s^{-1}.   427.   420 kV.

428. You should *explain* this use of the motor effect, noting its value in a case where very hot liquids are involved.

429. 270 m s^{-1}.

430. (a) 15 GJ; (b) 10 J.   431.   (a) 12 m^3; (b) 0.9 kg m^{-3}.

432. $m = 10\,000$ kg $g = 1.5$ m s^{-2} $W = ?$

$W = mg = 10\,000 \times 1.5$ N $= 15$ kN.

433. 10 Ω, 8.9Ω, 8 Ω, 6 Ω, 4 Ω, 2 Ω.

434. Z contains a lot of E, and a little each of D and G. F is absent.

435. Energy will pass from primary to secondary while the primary current grows from zero to its full value. After that there will be no transformer action – the laws of electromagnetic induction show zero secondary e.m.f. if the primary current is steady.

436. 40 μF.   437.   5 MNm^{-2}.

438. 1 V; 100 per cent.

439. $V = 5 \times 10^3$ V $d = 0.1$ m $E = ?$

$E = V/d = 5 \times 10^3/0.1$ V m^{-1} $= 50$ kV m^{-1} (or 50 kN C^{-1}).

440. Your essay should state the laws and relate each with the car. The ideas of force, velocity and acceleration should appear too.

441. $I = 2.5$ A $l = 0.05$ m $m = 0.025$ kg $a = 0.1$ m s^{-2} $B = ?$

$F = ma = 0.025 \times 0.1$ N $= 2.5 \times 10^{-3}$ N

$B = F/Il = 2.5 \times 10^{-3}/(2.5 \times 0.05)$ T $= 0.02$ T.

442. This requires you to follow through the relevant material in some detail in your own words.

443. 500 mm s^{-2}.

444. 1⅓ Ω.

445. Your essay should include a graph like those in Fig. 6.11, with all points well explained.

446. 96 N.

447. 170 nm, 1.6 Mm s^{-1}, 7.5 V.

448. 3.12×10^{-16} J; 4.72×10^{17} Hz; 6.36×10^{-10} m; gamma.

449. With $m = 1000$ kg $v_2 = 0$ m s^{-1} $v_1 = 10$ m s^{-1} we get $F = -10$ kN (a braking force).

450. Using $F = ma$ with $m = 5 \times 10^{-3}$ kg and $a = 10$ m s^{-2} gives 50 mN.

451. Chlorophyll absorbs green. Its spectrum is an absorption band covering the whole visible spectrum except for the green region.

452. This is based on the fact that $V_L \propto dI/dt$. The ideas are given in the text (section 17.3) but are worth looking at in detail.

453. 4 N.

454. 2.5C.

455. Your essay should expand on the nature of a fluid, the behaviour of its particles, and the friction between them as they flow. (See section 9.4.)

456. (a) 2.2×10^{-5} Hz; (b) 1.4×10^{-5} Hz; (c) (i) 15 μm s^{-2}, (ii) zero.

457. $Q = 5 \times 10^{-4}$ C $F = 10^3$ N $E = ?$

$E = F/Q = 10^3/(5 \times 10^{-4})$ N C^{-1} = 2MN C^{-1}

458. 18 protons and 22 neutrons (with 18 electrons in the cloud if the atom is neutral).

459. 20 nm.   **460.**   11.25 kN (the net acceleration is 7.5 m s^{-2}).

461. 400 kg.

462. The other values are 1.80×10^{-11} m, 1.85×10^{-11} m, 4.95×10^{-10} m, 1.98×10^{-9} m, 6.6×10^{-10} m. (Most are in fact X-ray quanta.)

463. 6.43×10^{13} Hz.   **464.**   1.2 MV m^{-1} or 1.2 MN C^{-1}.

465. 17 nm, 4.9 Mms^{-1}, 75 V.   **466.**   4 NC^{-1}; 40 N kg^{-1}.

467. There is one other isotope, nucleon number 63, about half as common as ^{65}Cu.

468. See text below and Fig. 1.5. **469.**   180 V.

470. The hydrogen atom is the simplest of all – it has only one electron in the cloud. Thus its energy level picture is only one series of allowed states; the line spectra produced follow changes up or down only that one ladder.

471. There is an image 12 mm on each side of the central one. The image nearest the slit is about eight times stronger than either of the others.

472. 27.

473. $W = (45/60) \times 2$ rads^{-1} = 4.7 rads^{-1}.

474. 125g.   **475.**   2 m.

476. *First electron*: $W = 3.9 \times 10^{-18}$ J $h = 6.6 \times 10^{-34}$ J s $c = 3.0 \times 10^8$ m s^{-1} $\lambda = ?$

$$W = h\nu \quad c = \nu\lambda$$

$$\rightarrow W = hc/\lambda \rightarrow \lambda = hc/W$$
$$= 6.6 \times 10^{-34} \times 3.0 \times 10^8/3.9 \times 10^{-18} \text{ m} = 51 \text{ nm}$$

Second electron: The wavelength is 23 nm.

477. See following text.

478. The capacitors should be in series.

479. (i) 12 J; (ii) 720 J.

480. The deflection will be 4/5 after the change – 64 mm.

481. Look with care at the unknown spectrum (d). Try to see if the lines of (a), (b) and (c) are all present in each case. The sample under test consists mainly of A, with a small content of C. B is absent. This is because the test spectrum (d) has bright (a) lines, faint (c) lines, but no (b) lines.

482. 2×10^{-4} s; 31 kHz.

483. Zero.

484. The relevant material of section 19.3 (with Fig. 19.4) should be discussed in your own words.

485. (*a*). In fact this value should be 8.2×10^{-19} J to be consistent with the other data.

486. See text following and Fig. 1.3.

487. (a) 10 kN C^{-1}; (b) 10 MN C^{-1}; (c) 50 MN C^{-1}.

488. The curve is like that of a diode.

489. Compare Fig. 12.13 and answers 481, 499, and 434.

490. 0.06 °C (0.06 K) (using $W_1 = mgh$, $W_2 = \eta W_1$, $W_2 = mc \triangle T$).

491. Your answer should proceed on rather the same lines as our example in the chapter, of the motion of a point on the edge of a record-player turntable.

492. You could use circuits as in Fig. 14.9. Expect to find a potential difference – current graph like that of Fig. 14.10(e) – both are solutions of chemical salts. At a given p.d. or current, the resistance would depend on the areas of the plates immersed in the solution, their distance apart and the concentration of the solution.

493. 1.3×10^8 m s^{-1}.

494. 10^{-2} m (10 mm).

495. The result is very much the same as that described for the magnetic field in Answer 231(a). Again the closeness of the field lines shows the field strength; the direction of the lines in this case is from positive to negative.

496. Nucleus: 92 protons and 143 neutrons; cloud of neutral atom: 92 electrons. This is 'uranium 235', a rare isotope of the metal, used in nuclear fission systems.

497. (b) 4.4 ms^{-1}, 009°.

498. 5.61×10^{13} Hz.

499. Your line spectrum should consist of heavy (b) lines and faint (a) lines.

500. (a) 500 kPa; (b)1 MPa; (c)200 K (–73 °C); (d) No – water vapour is not an 'ideal gas' at normal temperatures.

501. See Fig. 10.6 and accompanying text.

502. 12 A. **503.** No.

Index